Potential Flows

MECHANICAL ENGINEERING

A Series of Textbooks and Reference Books

1. Spring Designer's Handbook, *by Harold Carlson*
2. Computer-Aided Graphics and Design, *by Daniel L. Ryan*
3. Lubrication Fundamentals, *by J. George Wills*
4. Solar Engineering for Domestic Buildings, *by William A. Himmelman*
5. Applied Engineering Mechanics: Statics and Dynamics, *by G. Boothroyd and C. Poli*
6. Centrifugal Pump Clinic, *by Igor J. Karassik*
7. Computer-Aided Kinetics for Machine Design, *by Daniel L. Ryan*
8. Plastics Products Design Handbook, Part A: Materials and Components; Part B: Processes and Design for Processes, *edited by Edward Miller*
9. Turbomachinery: Basic Theory and Applications, *by Earl Logan, Jr.*
10. Vibrations of Shells and Plates, *by Werner Soedel*
11. Flat and Corrugated Diaphragm Design Handbook, *by Mario Di Giovanni*
12. Practical Stress Analysis in Engineering Design, *by Alexander Blake*
13. An Introduction to the Design and Behavior of Bolted Joints, *by John H. Bickford*
14. Optimal Engineering Design: Principles and Applications, *by James N. Siddall*
15. Spring Manufacturing Handbook, *by Harold Carlson*

16. Industrial Noise Control: Fundamentals and Applications, *edited by Lewis H. Bell*
17. Gears and Their Vibration: A Basic Approach to Understanding Gear Noise, *by J. Derek Smith*
18. Chains for Power Transmission and Material Handling: Design and Applications Handbook, *by the American Chain Association*
19. Corrosion and Corrosion Protection Handbook, *edited by Philip A. Schweitzer*
20. Gear Drive Systems: Design and Application, *by Peter Lynwander*
21. Controlling In-Plant Airborne Contaminants: Systems Design and Calculations, *by John D. Constance*
22. CAD/CAM Systems Planning and Implementation, *by Charles S. Knox*
23. Probabilistic Engineering Design: Principles and Applications, *by James N. Siddall*
24. Traction Drives: Selection and Application, *by Frederick W. Heilich III and Eugene E. Shube*
25. Finite Element Methods: An Introduction, *by Ronald L. Huston and Chris E. Passerello*
26. Mechanical Fastening of Plastics: An Engineering Handbook, *by Brayton Lincoln, Kenneth J. Gomes, and James F. Braden*
27. Lubrication in Practice, Second Edition, *edited by W. S. Robertson*
28. Principles of Automated Drafting, *by Daniel L. Ryan*
29. Practical Seal Design, *edited by Leonard J. Martini*
30. Engineering Documentation for CAD/CAM Applications, *by Charles S. Knox*
31. Design Dimensioning with Computer Graphics Applications, *by Jerome C. Lange*
32. Mechanism Analysis: Simplified Graphical and Analytical Techniques, *by Lyndon O. Barton*
33. CAD/CAM Systems: Justification, Implementation, Productivity Measurement, *by Edward J. Preston, George W. Crawford, and Mark E. Coticchia*
34. Steam Plant Calculations Manual, *by V. Ganapathy*
35. Design Assurance for Engineers and Managers, *by John A. Burgess*
36. Heat Transfer Fluids and Systems for Process and Energy Applications, *by Jasbir Singh*
37. Potential Flows: Computer Graphic Solutions, *by Robert H. Kirchhoff*

OTHER VOLUMES IN PREPARATION

Potential Flows

Computer Graphic Solutions

Robert H. Kirchhoff
Department of Mechanical Engineering
University of Massachusetts
Amherst, Massachusetts

MARCEL DEKKER, INC. New York and Basel

Library of Congress Cataloging in Publication Data

Kirchhoff, Robert H.,
Potential flows: computer graphic solutions

(Mechanical engineering; 37)
Includes index.
1. Fluid mechanics—Data processing. I. Title.
II. Series.
TA357.K57 1985 620.1'064'02854 84-25975
ISBN 0-8247-7307-1

MARCEL DEKKER, INC.
270 Madison Avenue, New York, New York 10016

Current printing (last digit):
10 9 8 7 6 5 4 3 2 1

PRINTED IN THE UNITED STATES OF AMERICA

Preface

This book is a compilation of 70 well-known potential flows in a useful graphical format. Dimensionless streamlines and isotachs plotted in the physical plane allow the solutions to be scaled to any application. The inclusion of isotachs is motivated by the belief that considerable physical insight into a fluid flow can be gained by studying these lines of constant speed. This collection is intended to be a reference for researchers and students in fluid mechanics. Extensive detailed derivations are not presented. The nomenclature of each solution is that of the original reference. The majority of references are readily available texts or reports.

All of the graphs were produced using an HP87 personal computer system linked to an HP7470A graphics plotter. Computer programs were written in BASIC utilizing the Hewlett Packard Graphics Language.

Many people are to be thanked for their contribution to this book. Mr. Kendall Miller, a former student, introduced me to computer graphics. Dr. David Pratt, Chairman of the Mechanical Engineering Department at the University of

Washington, provided a sabbatical haven where I started the book. Dr. J. R. Dixon and Dr. D. E. Cromack in their tenure as Chairmen of the Mechanical Engineering Department at the University of Massachusetts helped me in numerous ways to bring the project to completion. Mr. Dan Terpack and Mr. Rick Connolly of the Hewlett Packard Corp. arranged for the loan of computer software and firmware which was essential to the programming effort.

I hope that this collection of graphical solutions will promote understanding and use of these potential flows. I extend an invitation to users of this book to suggest flows not included herein for consideration in future editions of this collection.

Robert H. Kirchhoff

Contents

List of Figures

List of Figures

List of Figures

Potential Flows

1
Two Dimensional Flows

1.1 INTRODUCTION

The first chapter of this book is a graphical exposition of two dimensional incompressible potential flow of a frictionless fluid. These very special flows are key building blocks for the understanding of more complicated fluid motion. The solutions are drawn primarily from Milne-Thomson (1960), Streeter (1948), Currie (1974) and Kennard (1967). The notation of each of these sources is very similar; thus, no attempt was made to unify it.

All of the flows depicted in this book are for a fluid of zero viscosity. This is equivalent to saying that the fluid cannot sustain a shear stress or that the fluid is frictionless. Such flows have also been referred to by Feynman (1964) as "the flow of dry water." The graphical solutions are presented in terms of dimensionless plots of streamlines and isotachs. The following brief exposition of two dimensional potential flow theory should serve as a guide for interpreting the computer graphic plots.

Potential Flows

The first governing equation of fluid motion expressing the conservation of mass is known as the continuity equation. In two dimensional motion for fluids of constant density ρ, the continuity equation takes the form

$$\frac{\partial u}{\partial x} + \frac{\partial v}{\partial y} = 0$$

where u and v are the x and y components of velocity respectively. The stream function $\psi(x,y)$ defined as

$$u = -\frac{\partial \psi}{\partial y}, \quad v = \frac{\partial \psi}{\partial x}$$

is a solution to the continuity equation. Thus, streamlines have a variety of interpretations associated with the conservation of mass in fluid flow.

The second governing equation of fluid motion, expressing the conservation of momentum, is the Euler equation. For the steady flow of a frictionless fluid of constant density in the absence of body forces, the x and y components of the Euler equation are written

$$u \frac{\partial u}{\partial x} + v \frac{\partial u}{\partial y} = -\frac{1}{\rho} \frac{\partial p}{\partial x}$$

$$u \frac{\partial v}{\partial x} + v \frac{\partial v}{\partial y} = -\frac{1}{\rho} \frac{\partial p}{\partial y}$$

where $p(x,y)$ is the pressure, ρ is the constant density, and u,v are the x and y velocity components. The continuity equation plus the two Euler equations make a closed set of three equations in three unknowns u,v,p. In the absence of viscosity, the fluid does not stick to a fixed boundary. Thus, the only boundary condition is that the fluid have no velocity component normal to the surface. This boundary condition is expressed as

$$\ell u + mv = 0$$

where ℓ and m are the direction cosines of the normal to the surface. For a curl free two dimensional velocity field expressed as

$$\frac{\partial v}{\partial x} - \frac{\partial u}{\partial y} = 0$$

the velocity may be written as the gradient of a scalar potential $\phi(x,y)$

$$u = -\frac{\partial \phi}{\partial x}, \quad v = -\frac{\partial \phi}{\partial y}$$

Under such conditions, the flow is known as irrotational potential flow. Curl free fluid motion may be further interpreted by saying that in irrotational potential flow, the fluid motion is comprised of translation plus pure strain. Motion generated by a ship or an aircraft moving through a fluid is roughly irrotational except in the thin boundary layer near the solid surface where friction and viscosity play a dominant role. All the fluid motions depicted in this book are for irrotational flow outside the boundary layer.

Substituting the definition of the velocity potential into the continuity equation yields

$$\nabla^2 \phi = 0$$

and substituting the definition of the stream function into the curl free expression yields

$$\nabla^2 \psi = 0$$

Thus, ϕ and ψ form an orthogonal set of lines in the potential flow known as the flow net. If either the velocity potential or the stream function is known, the flow is fully specified in terms of pressure and velocity through the Euler equations. Lines of ϕ = constant are equipotential lines. The velocity is normal to the equipotential lines and the sign of the velocity potential is chosen such that the potential decreases in the direction of the velocity. Lines of ψ = constant are streamlines which are everywhere parallel to the velocity vector.

Euler's equations may be integrated into the Bernoulli equation which for steady irrotational motion of a fluid of constant density is written

$$\frac{p}{\rho} + \frac{q^2}{2} = \text{constant}$$

where $q^2 = u^2 + v^2$ and the constant has the same value on all the streamlines.

Since the presentation in this book relies extensively on the stream function, the following properties are highlighted:

1. The stream function is a solution to the continuity equation.
2. The value of ψ is constant along a streamline.
3. The flow is parallel to the streamlines and thus any line of ψ = constant may be interpreted as a solid boundary.
4. Streamlines are concave toward the side on which the magnitude of the velocity is larger.
5. The velocity increases in the direction in which the streamlines converge and, conversely, the velocity decreases in the direction in which the streamlines diverge.
6. ψ = constant on any stationary boundary and this constant is usually chosen to be zero.

7. The volume flow between any two streamlines is equal to the difference in the value of their stream functions, ψ.

8. The stream function ψ has the units of $(\text{length})^2/\text{time}$.

In two-dimensional hydrodynamics, extensive use is made of the functions of a complex variable. The independent variables x,y are combined into the complex variable $z = x + iy$. The dependent variables ϕ and ψ are combined into a variable known as the complex potential w.

$$w = \phi + i\psi$$

The development of most of the plots starts with the complex potential w. The stream function is usually obtained by writing

$$2i\psi = w(z) - \overline{w}(\overline{z})$$

The velocity is related to the complex potential through the complex velocity function

$$\frac{dw}{dz} = -u + iv$$

The speed of the flow is the magnitude of the velocity

$$q = (u^2 + v^2)^{1/2}$$

In practice, the speed is usually obtained from the expression

$$q^2 = \frac{dw}{dz}\frac{d\overline{w}}{d\overline{z}}$$

Lines of constant speed q are called isotachs.

In all cases, the resulting streamlines and isotachs are plotted as dimensionless quantities PSI and Q respectively. In most cases, the streamlines are non-dimensionalized by a length a and a speed U as PSI = ψ/Ua and the isotachs are non-dimensionalized by a speed U and Q = q/U. In only the simplest cases was it possible to directly solve for x = f(y,PSI) and x = g(y,Q). The most common method of solution was to develop an equation for either the streamlines or isotachs as PSI = f(x,y) or Q = Q(x,y) and then use a Newton-Raphson technique to invert them for plotting.

A few sample programs presented in the Appendix will give the reader the flavor of the solution techniques and the software development.

The natural symmetry of these flows has often been utilized to separate the streamlines and the isotachs along the PSI = 0 line. Local velocity vectors can be constructed by overlaying a transparency of the isotachs onto the streamlines, the direction of the speed being parallel to the streamlines.

Further interpretation of these flows can be made by comparing them to a number of the examples in the recent collection of photographs of fluid flows assembled by Van Dyke (1982).

The potential flow of incompressible fluids has analogies to other fields. For example, the fluid velocity is analogous to the electrostatic field strength, the magnetizing force and the flux of heat. These analogies are explored in Weber (1950) and Fuller (1973). A number of the plots presented herein are helpful in understanding these other fields of physics.

1.2 FLOW IN A CORNER

Milne-Thomson (1960), page 151, gives the complex potential for flow in a corner of angle α.

$$w = Ua\left(\frac{z}{a}\right)^{\pi/\alpha} \tag{1}$$

The stream function ψ may be written from Equation (1) as

$$\psi = Ua\left(\frac{r}{a}\right)^{\pi/\alpha} \sin\frac{\pi\theta}{\alpha} \tag{2}$$

The dimensionless stream function PSI is defined as PSI = ψ/Ua and equation (2) is inverted to generate lines of constant PSI as

$$\frac{r}{a} = \left(\frac{PSI}{\sin\frac{\pi\theta}{\alpha}}\right)^{\alpha/\pi} \tag{3}$$

Differentiating Equation (1) to formulate the complex velocity dw/dz and the speed q yields

$$\frac{q}{U} = \frac{\pi}{\alpha}\left(\frac{r}{a}\right)^{\beta} \tag{4}$$

where $\beta = (\pi/\alpha) - 1$. The dimensionless isotachs are lines of constant Q where Q = q/U. Equation (4) is directly inverted to generate the isotachs.

$$\frac{r}{a} = \left(\frac{\alpha}{\pi}\, Q\right)^{1/\beta} \tag{5}$$

Figures 1, 2, 3, and 4 show the dimensionless streamlines and isotachs for angles $\alpha = \pi/4$, $\pi/2$, $3\pi/2$ and 2π respectively. The flow stagnates in a corner of $\alpha \leqslant \pi/2$ but the speed goes to infinity in a corner of $\alpha > \pi/2$.

The length a and the speed U are not "natural" scales of this problem; they merely act as convenient parameters to formulate dimensionless PSI and Q.

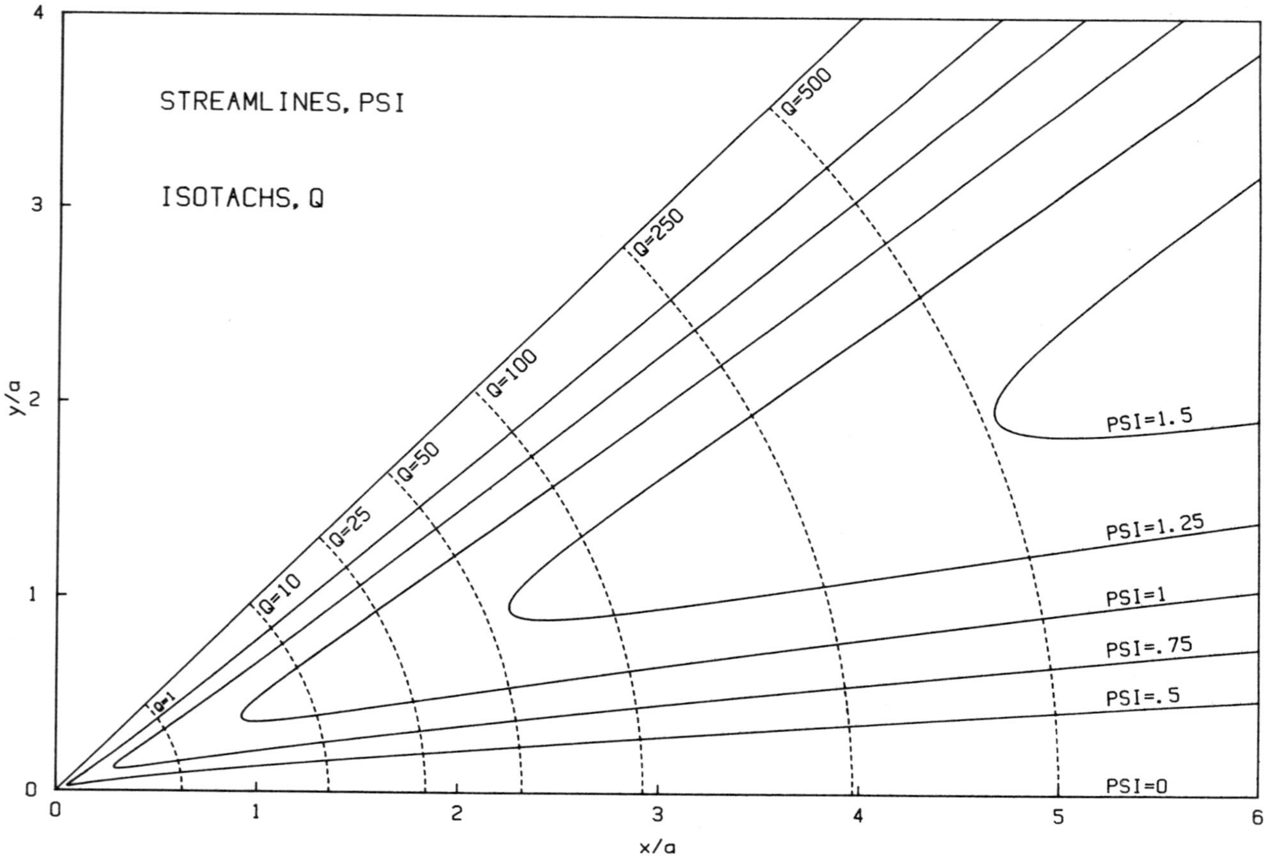

Figure 1 Flow in a 45 degree corner.

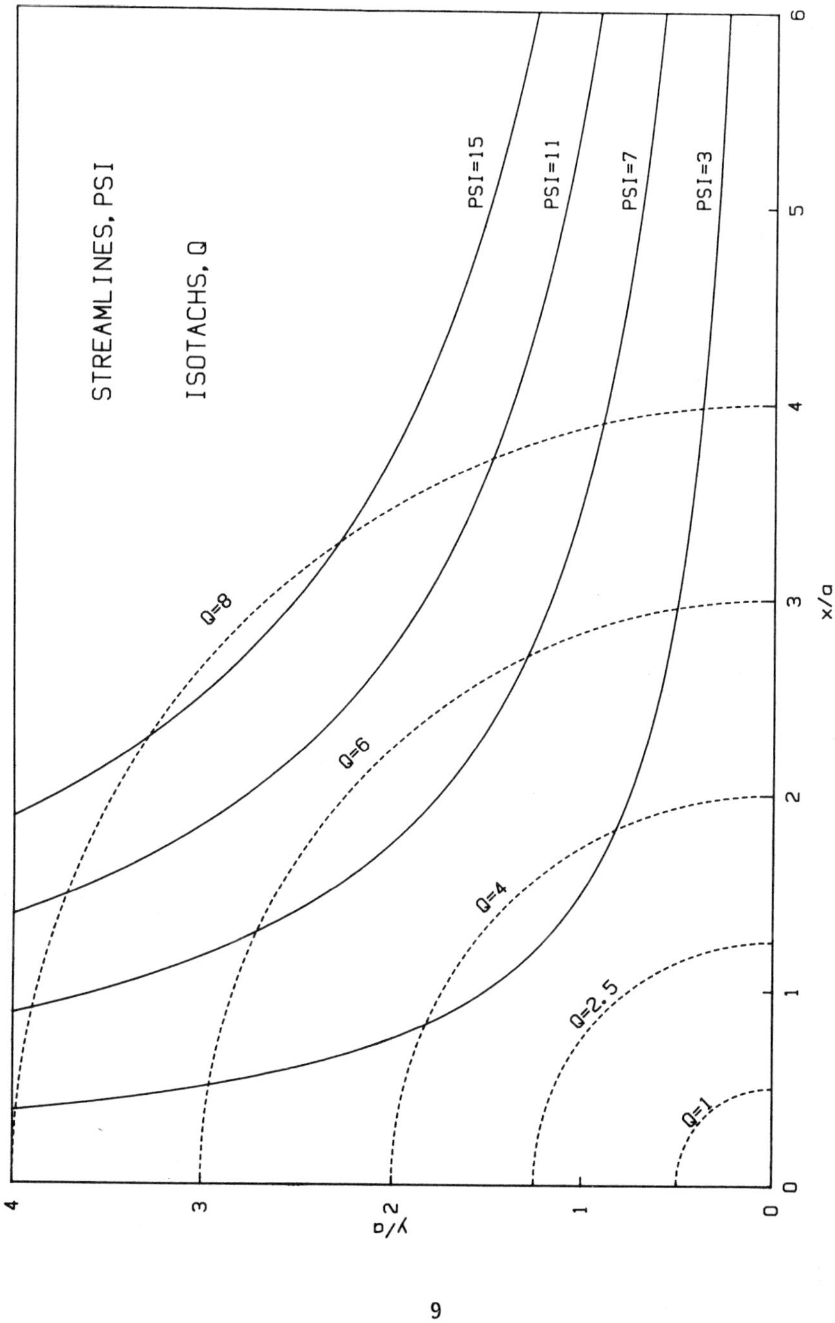

Figure 2 Flow in a 90 degree corner.

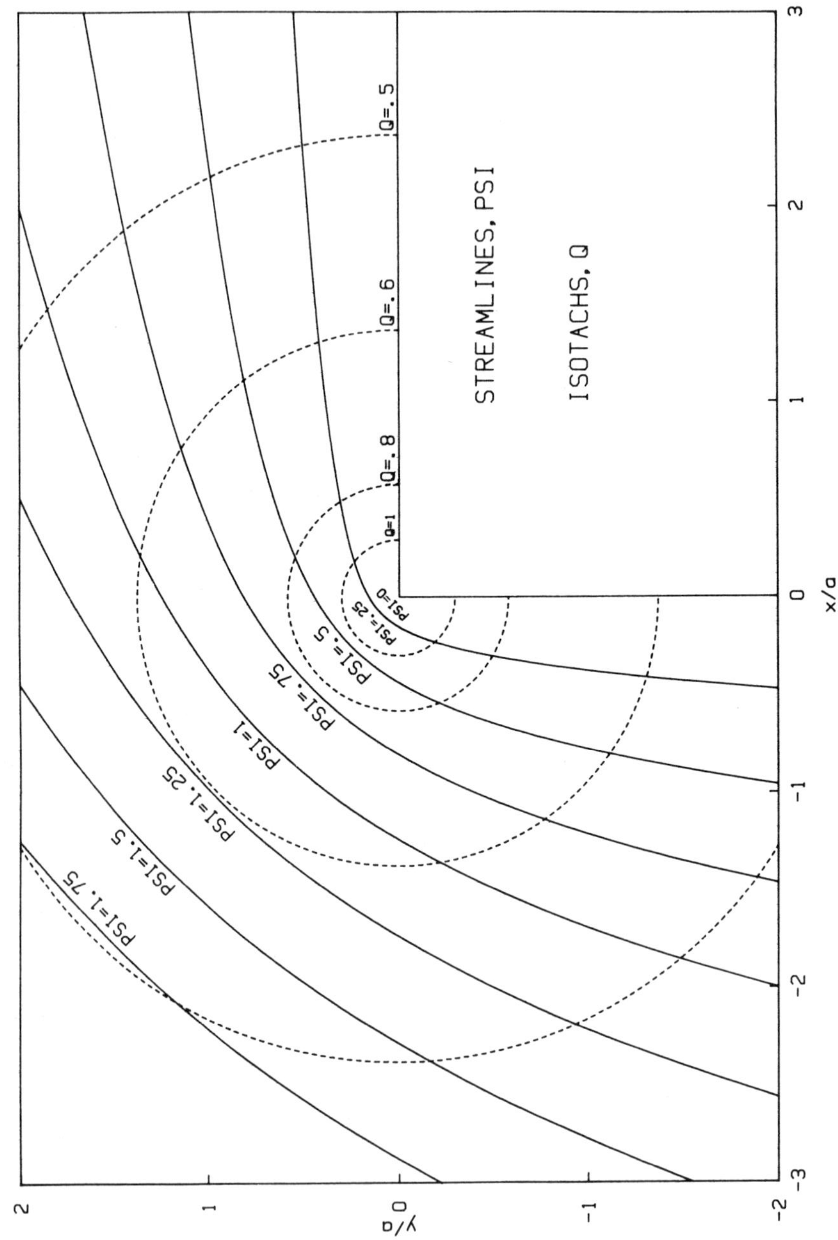

Figure 3 Flow in a 270 degree corner.

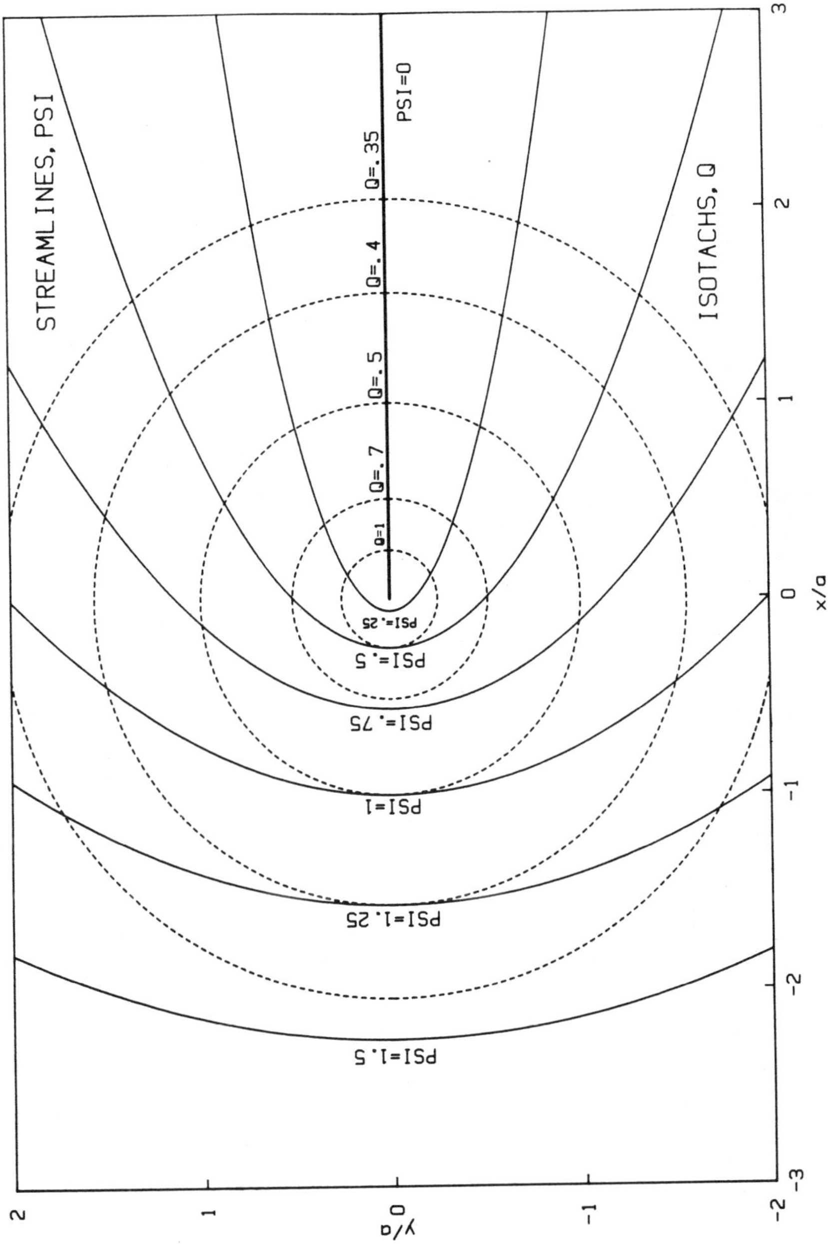

Figure 4 Flow over a 360 degree corner.

1.3 A SOURCE

Milne-Thomson (1960), page 201, gives the complex potential for a simple source as

$$w = -m \log z \tag{6}$$

The stream function is $\psi = -m\theta$, and the dimensionless stream function is $PSI = \psi/m$. Thus, the equation to be plotted is

$$\theta = -PSI \tag{7}$$

Differentiating equation (6) to formulate the complex potential and the speed q yields $q = m/r$. The dimensionless isotachs are defined as $Q = qa/m$ and the equation to be plotted is

$$r/a = 1/Q \tag{8}$$

Dimensionless streamlines and isotachs are plotted on Figure 5.

1.4 A DOUBLET

Milne-Thomson (1960), page 204, gives the complex potential for the doublet of strength μ at angle α to the x axis as

$$w = \frac{\mu e^{i\alpha}}{z} \tag{9}$$

When $\alpha = 0$, the equation for the streamlines may be written, suitably nondimensionlized by length scale a, as

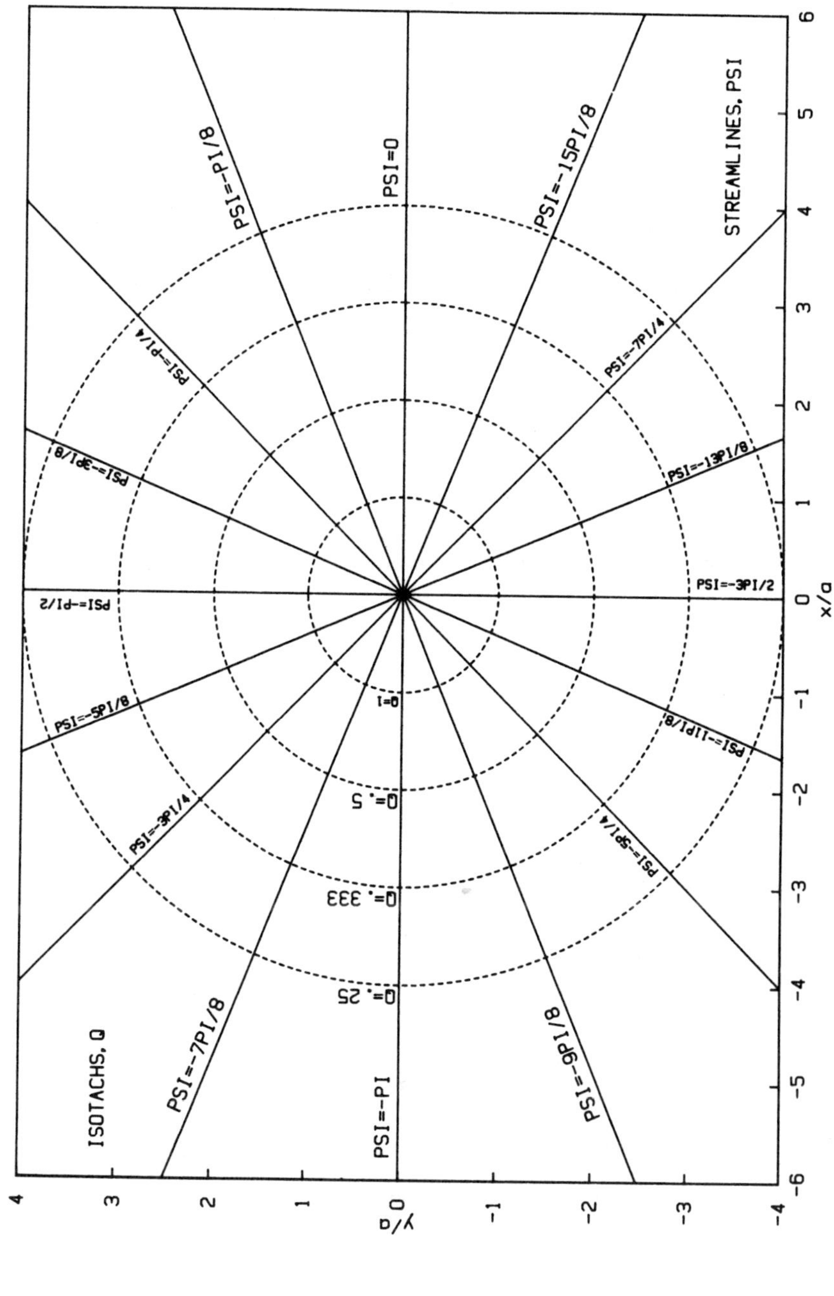

Figure 5 A simple source.

$$\frac{x^2}{a^2} + \left(\frac{y}{a} + \frac{\mu/a}{2\psi}\right)^2 = \left(\frac{\mu/a}{2\psi}\right)^2 \tag{10}$$

From equation (10) the dimensionless stream function is PSI = $\psi a/\mu$. Streamlines are circles tangent to the x axis as seen on Figure 6. The doublet has a natural symmetry about the y axis; thus, the line x = o on Figure 6 is not part of the flow but merely serves to separate the streamlines and the isotachs.

Differentiating equation (9) to formulate dw/dz and then q results in

$$\frac{qa^2}{\mu} = \left(\frac{a}{r}\right)^2 \tag{11}$$

The dimensionless isotachs are defined as $Q = qa^2/\mu$ and are plotted as circles on Figure 6.

1.5 A SOURCE AND A SINK

Milne-Thomson (1960), page 204, gives the complex potential for a source at x = a and a sink at x = -a as

$$w = -m \log (z - a) + m \log (z + a) \tag{12}$$

The equation for the streamlines developed from equation (12), suitably nondimensionalized by the length a, may be written as

$$\left(\frac{x}{a}\right)^2 + \left(\frac{y}{a} - \cot\left(-\frac{\psi}{m}\right)\right)^2 = 1 + \cot^2\left(-\frac{\psi}{m}\right) \tag{13}$$

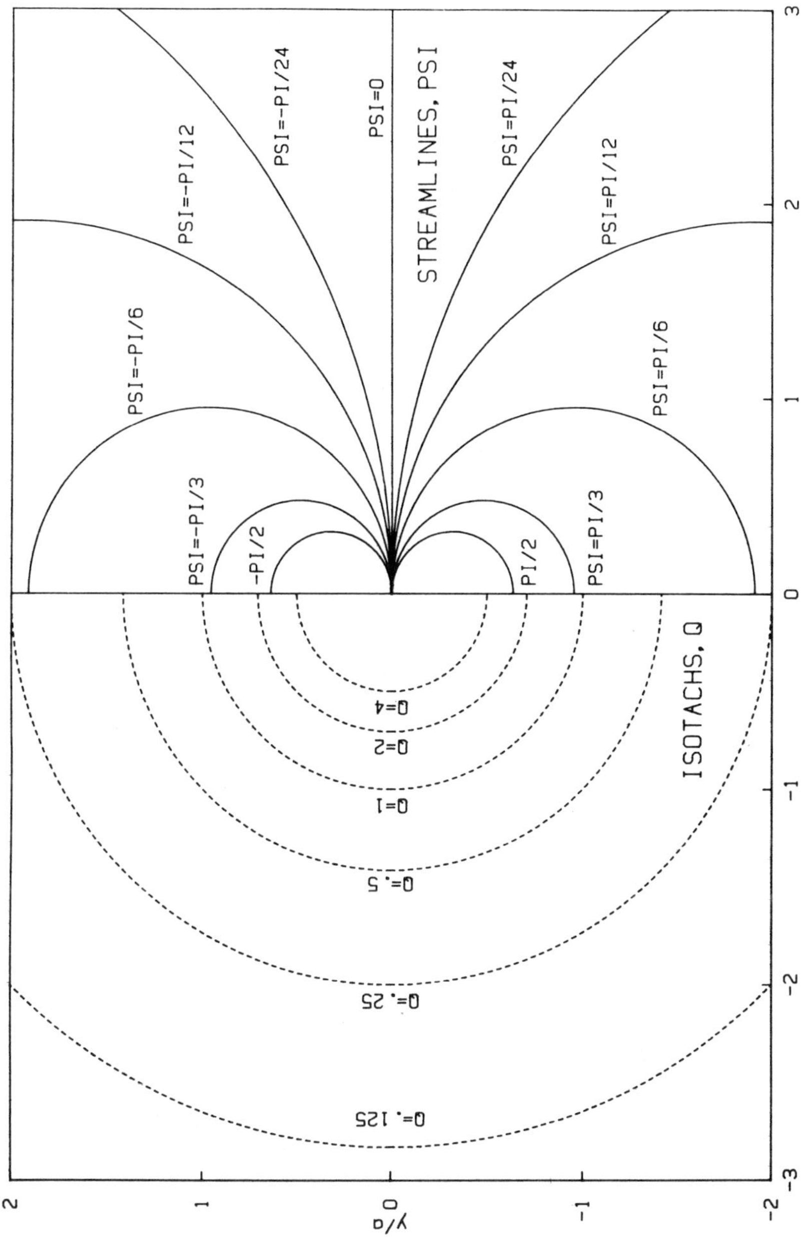

Figure 6 A doublet at the origin.

The dimensionless stream function is PSI = ψ/m and the streamlines are plotted as circles from equation (13) on Figure 7. The line y = 0, also PSI = 0, provides a natural line of symmetry separating the streamlines and isotachs.

Differentiating equation (12) to formulate the complex velocity and then the speed q gives

$$\frac{q^2 a^2}{m^2} = \frac{4}{\left(\frac{z^2}{a^2} - 1\right)\left(\frac{\bar{z}^2}{a^2} - 1\right)} \tag{14}$$

The dimensionless isotachs are defined as Q = qa/m. When equation (14) is expanded, the isotachs are represented by a fourth order polynomial in x and y which cannot be directly solved for x as a function of y and Q. The isotachs on Figure 7 were obtained by solving equation (14) using a standard Newton-Raphson method (James, Smith and Wolford, 1967).

1.6 A SOURCE AND A SINK IN A FLOW

Milne-Thomson (1960), page 206, gives the complex potential for a source at x = a and a sink at x = -a in a uniform stream U flowing from right to left.

$$w = Uz - m \log (z - a) + m \log (z + a) \tag{15}$$

When the dimensionless stream function is chosen to be PSI = ψ/Ua and the dimensionless source strength is m/Ua = 1, the streamline equation can be developed from equation (15) as

$$\left(\frac{x}{a}\right)^2 = 1 - \left(\frac{y}{a}\right)^2 + \frac{2(y/a)}{\tan\left(\frac{y}{a} - \mathrm{PSI}\right)} \tag{16}$$

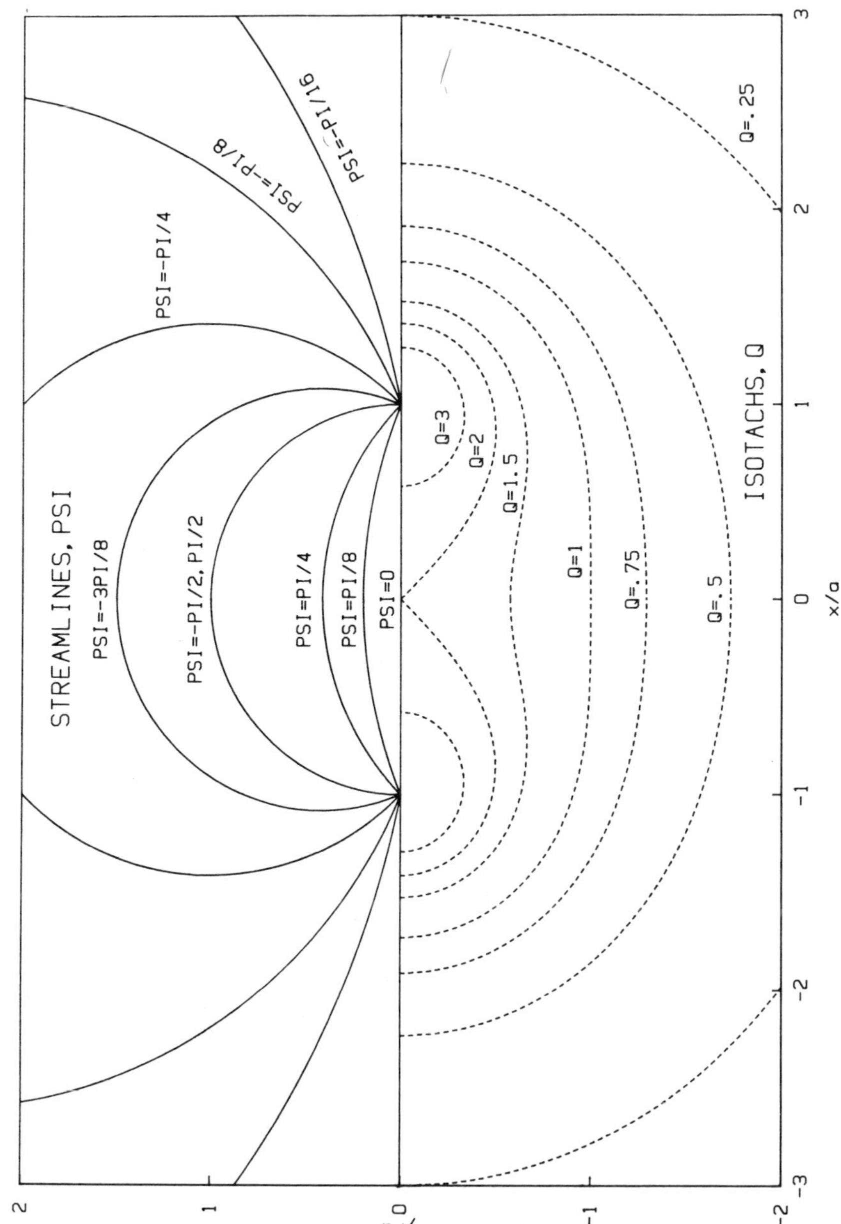

Figure 7 A source at x/a = 1 and a sink at x/a = 1.

The streamline PSI = 0 in equation (16) forms a solid body shape which intersects the x axis at $x/a = \pm (3)^{1/2}$ and intersects the y axis at $y/a = \pm 1.3065$. The aspect ratio of the solid body is varied by changing the dimensionless source strength m/Ua. Streamlines for PSI $\geqslant 0$ from equation (16) are shown in the upper half of Figure 8.

The complex velocity and the speed q can be formulated from equation (15), and when the dimensionless isotachs are defined to be lines of $Q = q/U$, then the equation for the isotachs in polar coordinates (r, θ) is

$$\cos 2\theta = \frac{(9 - Q^2) + (1 - Q^2)(r/a)^4}{(6 - 2Q^2)(r/a)^2} \tag{17}$$

The isotachs from equation (17) are plotted on Figure 8 with stagnation points at $x/a = \pm (3)^{1/2}$ and the maximum value of $Q = 1.7388$ at $y/a = \pm 1.3065$.

1.7 A SOURCE IN A UNIFORM STREAM

Eskinazi (1967), page 303, gives the stream function in polar coordinates for a source of strength K in a uniform stream U flowing from right to left as

$$\psi = -\frac{K}{2\pi}\theta + U r \sin\theta \tag{18}$$

The plane half body thus formed has a nose radius at $\theta = 0$ of $R = K/2\pi U$. When the dimensionless stream function is PSI = ψ/UR; equation (18) may be written in dimensionless form as

$$\frac{x}{R} = \frac{y/R}{\tan(y/R - \mathrm{PSI})} \tag{19}$$

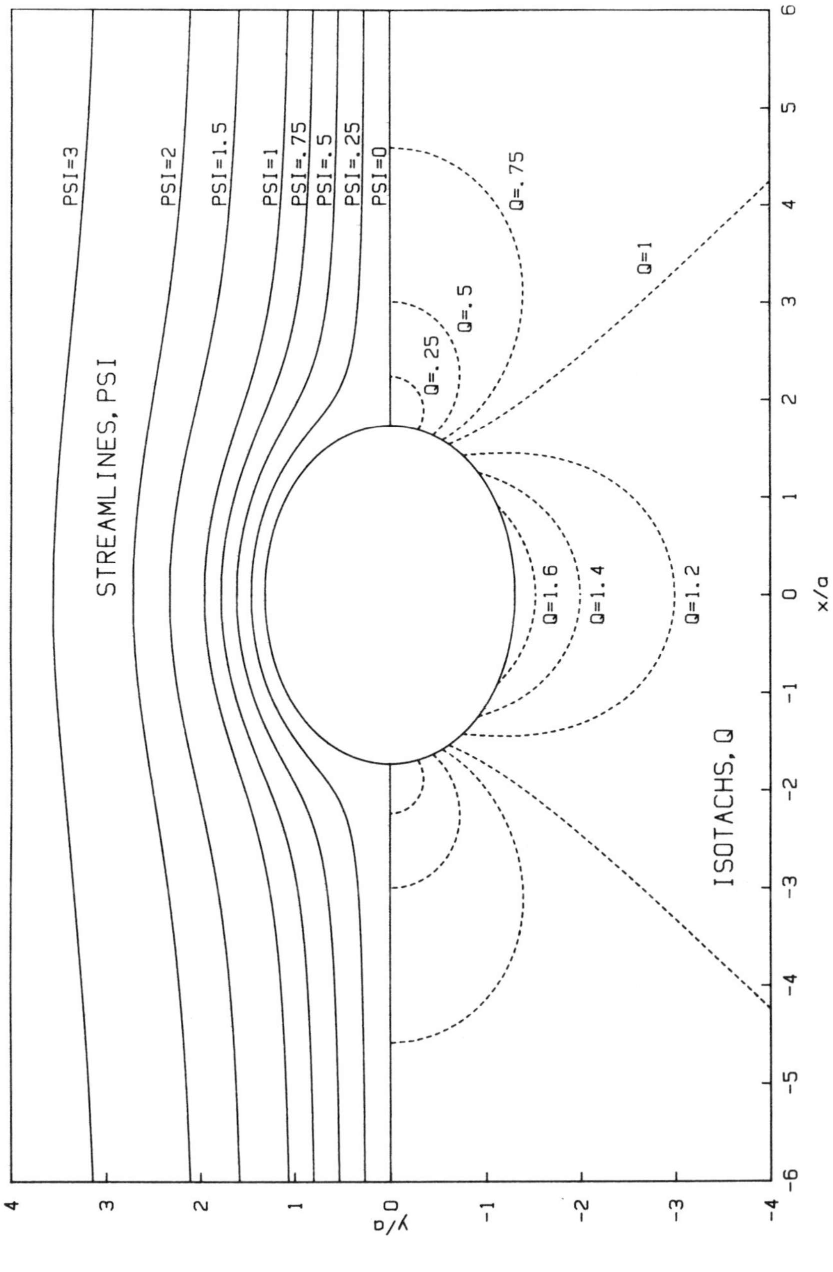

Figure 8 Equal source and a sink in a stream.

Streamlines PSI $\geqslant 0$ and the halfbody are plotted on the Figure 9. The asympototic height of the halfbody as $\theta \to \pi$ is $y/R = \pi$.

The velocity components u and v in the x and y directions are also given by Eskinazi (1967). When the dimensionless speed is $Q = q/U$, the equation for the isotachs may be written

$$(1 - Q^2)(r/R)^2 - 2 \cos \theta \, (r/R) + 1 = 0 \qquad (20)$$

Isotachs from equation (20) are plotted on Figure 9. The flow stagnates at $\theta = 0$ on the body, then accelerates to $Q = 1$ at $\theta = 67°$ and further accelerates to $Q = 1.25$ at $\theta = 117°$. From here, the flow decelerates as it approaches $Q = 1$ as $\theta \to \pi$.

1.8 TWO EQUAL SOURCES

Milne-Thomson (1960), page 208, gives the complex potential for two equal sources of strength m, one at $x = a$ and the other at $x = -a$.

$$w = -m \log (z - a) - m \log (z + a) \qquad (21)$$

This combination of a source and its image may also be thought of a the flow due to a source near a solid wall. When the dimensionless stream function is defined to be PSI $= \psi/m$; the equation of the streamlines may be written from equation (21) as the quadratic

$$\left(\frac{x}{a}\right)^2 + 2\left(\frac{x}{a}\right)\left(\frac{y}{a}\right) \cot (\text{PSI}) - \left(\frac{y}{a}\right)^2 = 1 \qquad (22)$$

The streamlines for $0 \leqslant \text{PSI} \leqslant \pi$ for the source at $x = a$ are plotted on Figure 10. There is an identical set of streamlines $\pi \leqslant \text{PSI} \leqslant 2\pi$ from the source at $x = -a$ which are not shown

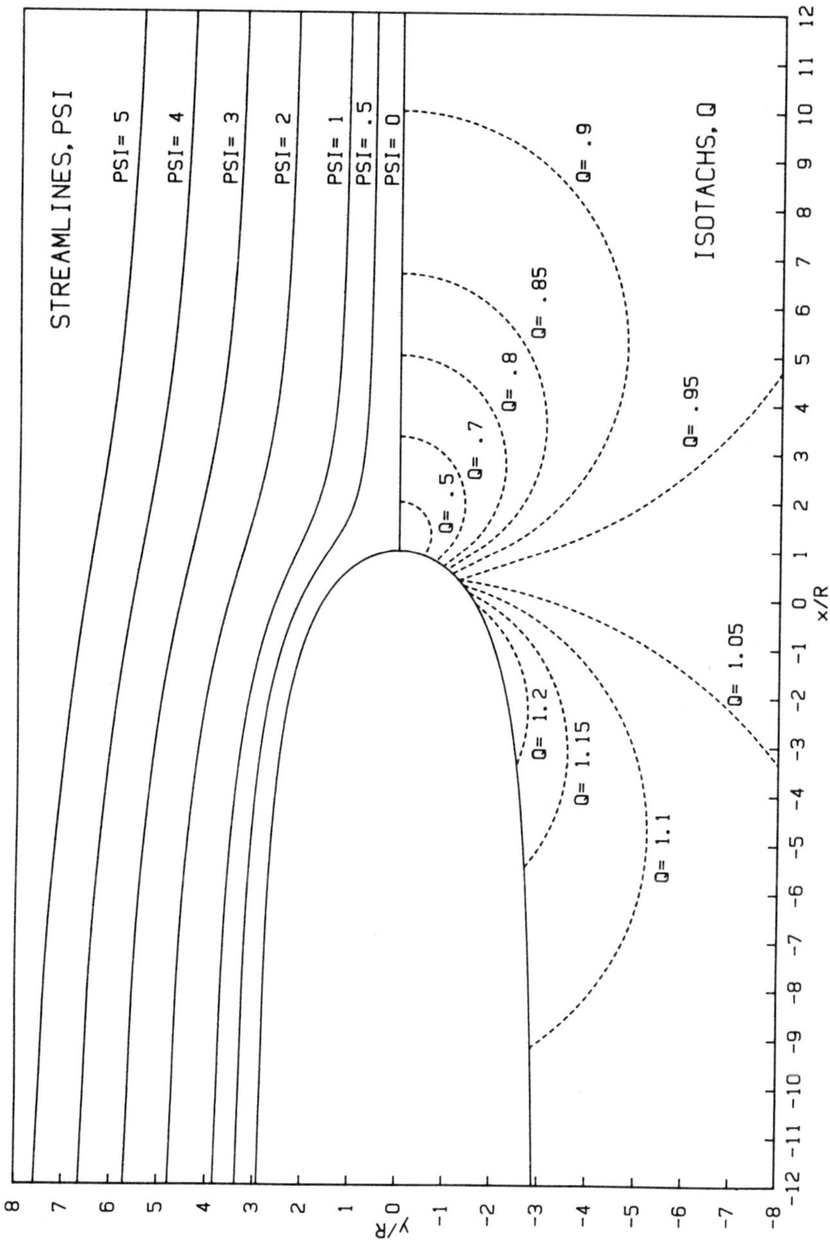

Figure 9 A source in a uniform stream.

Equation (21) can be differentiated to form the complex velocity and the speed q as

$$\frac{q^2 a^2}{m^2} = \frac{4z\bar{z}}{\left(\frac{z^2}{a^2} - 1\right)\left(\frac{\bar{z}^2}{a^2} - 1\right)} \tag{23}$$

The dimensionless speed was chosen to be $Q = qa/m$ and the resulting fourth order polynomial for the isotachs from equation (23) was solved numerically using a Newton Raphson technique (James, Smith and Wolford, 1967). The isotachs are plotted on the left half of Figure 10. Note the maximum of $Q = 1$ on the wall at $y = \pm a$.

1.9 Source Between Two Walls

Streeter (1948), page 112, gives the complex potential for a source between two walls separated by a distance a

$$w = c \ln \sinh \frac{\pi z}{a} \tag{24}$$

When the constant c is chosen to be $c = Ua/\pi$, then the speed of the flow far from the source is $q = U$. The dimensionless stream function is chosen to be $PSI = \psi/Ua$ which allows the equation for the streamlines to be developed from (24) as

$$\frac{y}{a} = \frac{1}{\pi} \tan^{-1}\left(\tanh\left(\frac{\pi x}{a}\right)\tan(\pi PSI)\right) \tag{25}$$

Streamlines are shown on the upper half of Figure 11. Notice the discontinuity across the vertical streamline.

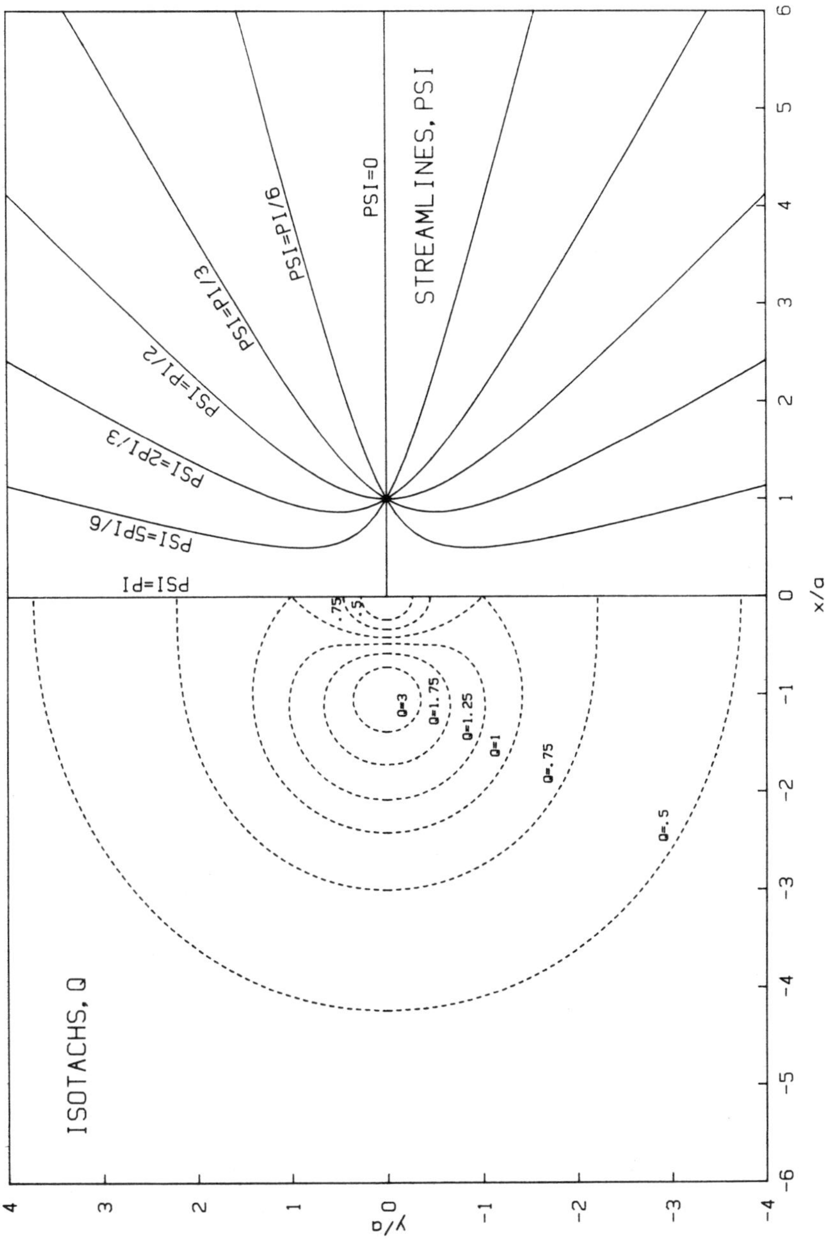

Figure 10 A source and its image.

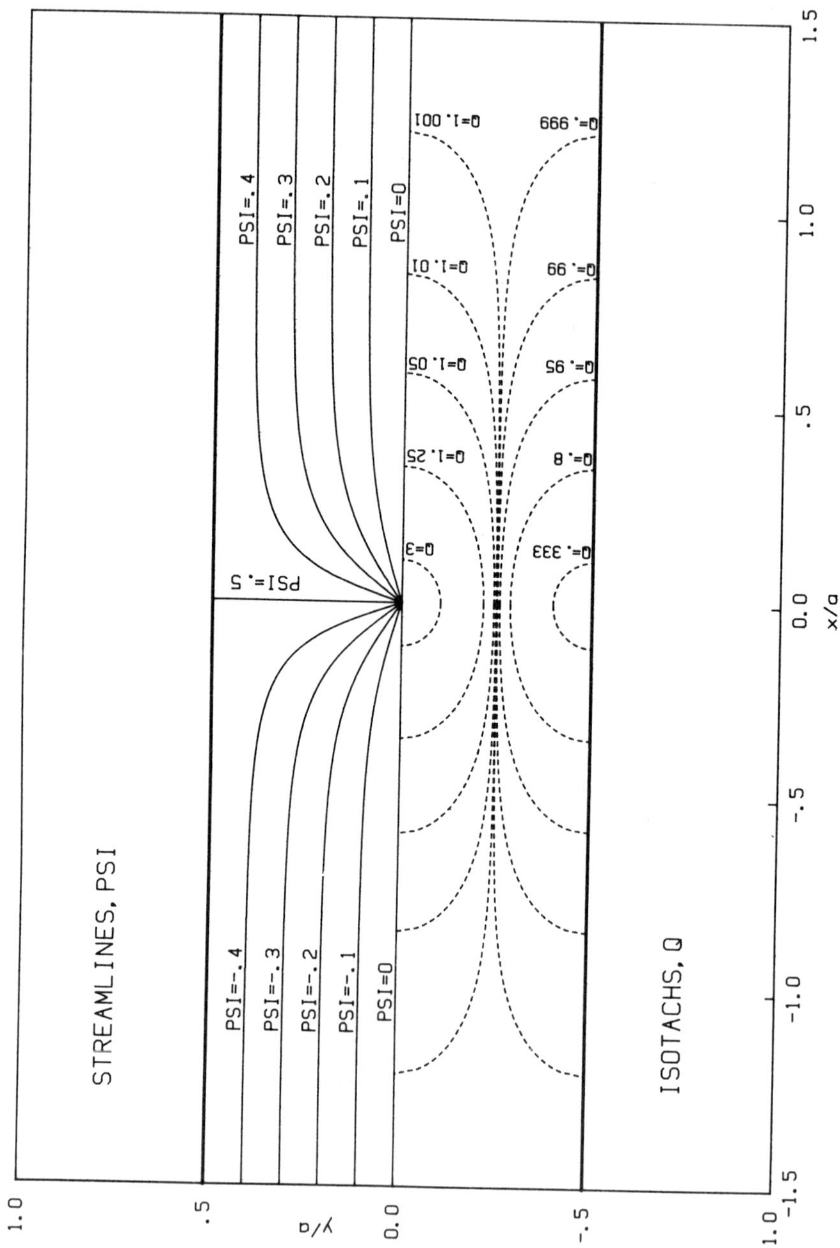

Figure 11 A source between two walls.

The dimensionless isotachs are defined to be $Q = q/U$. When (24) is differentiated to obtain the complex velocity dw/dz and the speed q, the dimensionless isotach equation may be written

$$Q^2 = \frac{\sinh^2(\pi x/a) + \cos^2(\pi y/a)}{\sinh^2(\pi x/a) + \sin^2(\pi y/a)} \tag{26}$$

Equation (26) can be directly inverted to solve for $y = f(x,Q)$ and the result is plotted on the lower half of Figure 11. Notice the rapid adjustment to uniform flow. In less than one wall spacing, the flow is within 1 percent of uniform flow.

1.10 FLOW OVER A CYLINDER

Milne-Thomson (1960), page 154, gives the complex potential for a cylinder of radius a in a uniform flow from right to left

$$w = U\left(z + \frac{a^2}{z}\right) \tag{27}$$

The dimensionless stream function is $PSI = \psi/Ua$ and the equation for the streamlines can be written from (27) in dimensionless form

$$\sin\,\theta = \frac{PSI}{\left(\frac{r}{a} - \frac{a}{r}\right)} \tag{28}$$

Streamlines for the flow are shown in the upper half of Figure 12.

Differentiating (27) to formulate dw/dz yields the isotach equation

$$\cos 2\theta = \frac{\left(\frac{a}{r}\right)^4 + 1 - Q^2}{2\left(\frac{a}{r}\right)^2} \tag{29}$$

where $Q = q/U$. Isotachs are plotted in the lower half of Figure 12. The flow stagnates at $x = a$, $y = 0$, then accelerates to $Q = 1$ at $\theta = 30°$ on the cylinder and further accelerates to a maximum of $Q = 2$ at $\theta = \pi/2$.

1.11 A TRANSLATING CYLINDER

Milne-Thomson (1960), page 234, gives the complex potential for a cylinder of radius a moving at speed U into a liquid at rest.

$$w = \frac{Ua^2}{z} \tag{30}$$

Equation (30) is identical with the potential for a doublet at the origin already treated in Section 1.4. The dimensionless stream function is $PSI = \psi/Ua$ and the streamline equation is developed from (30) as

$$\frac{r}{a} = \frac{-\sin\theta}{PSI} \tag{31}$$

Equation 30 is differentiated to form the complex velocity and the speed resulting in

$$Q = \frac{a}{r} \tag{32}$$

where $Q = q/U$.

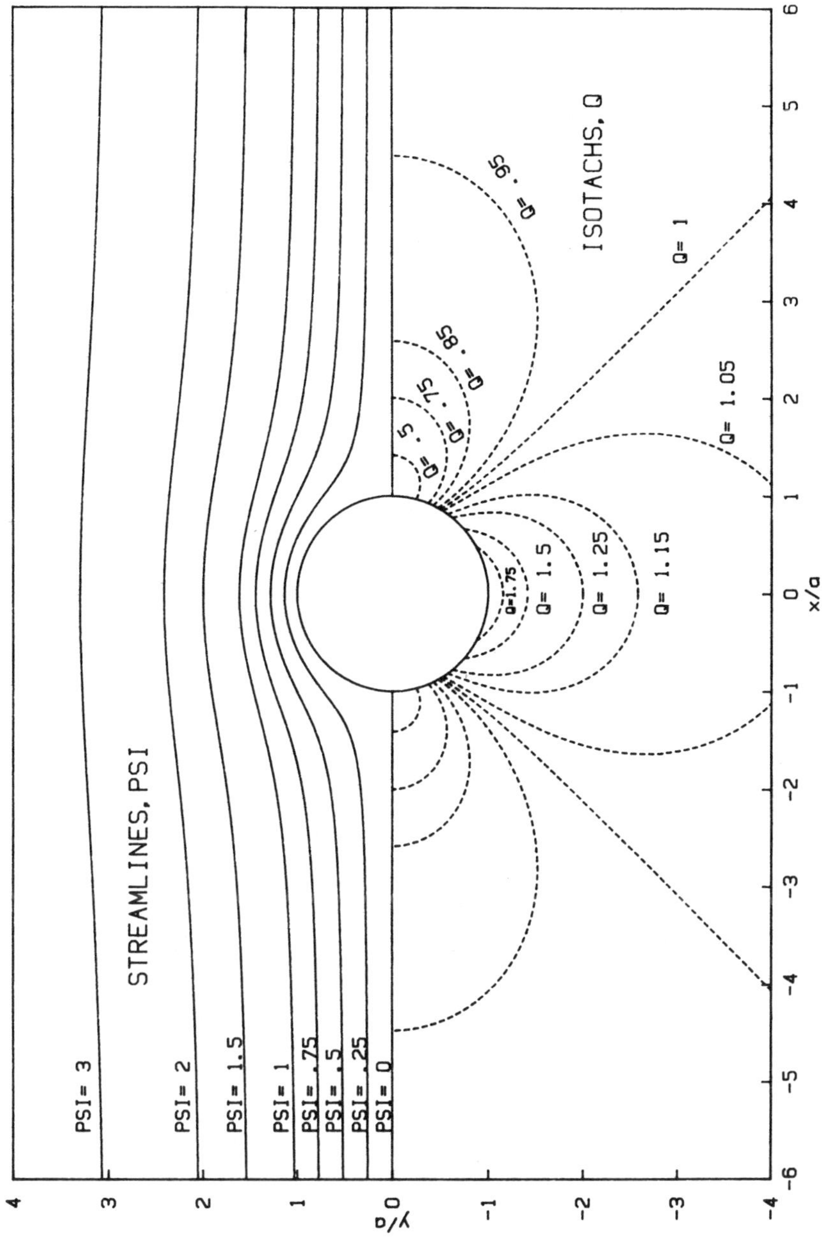

Figure 12 Flow over a cylinder.

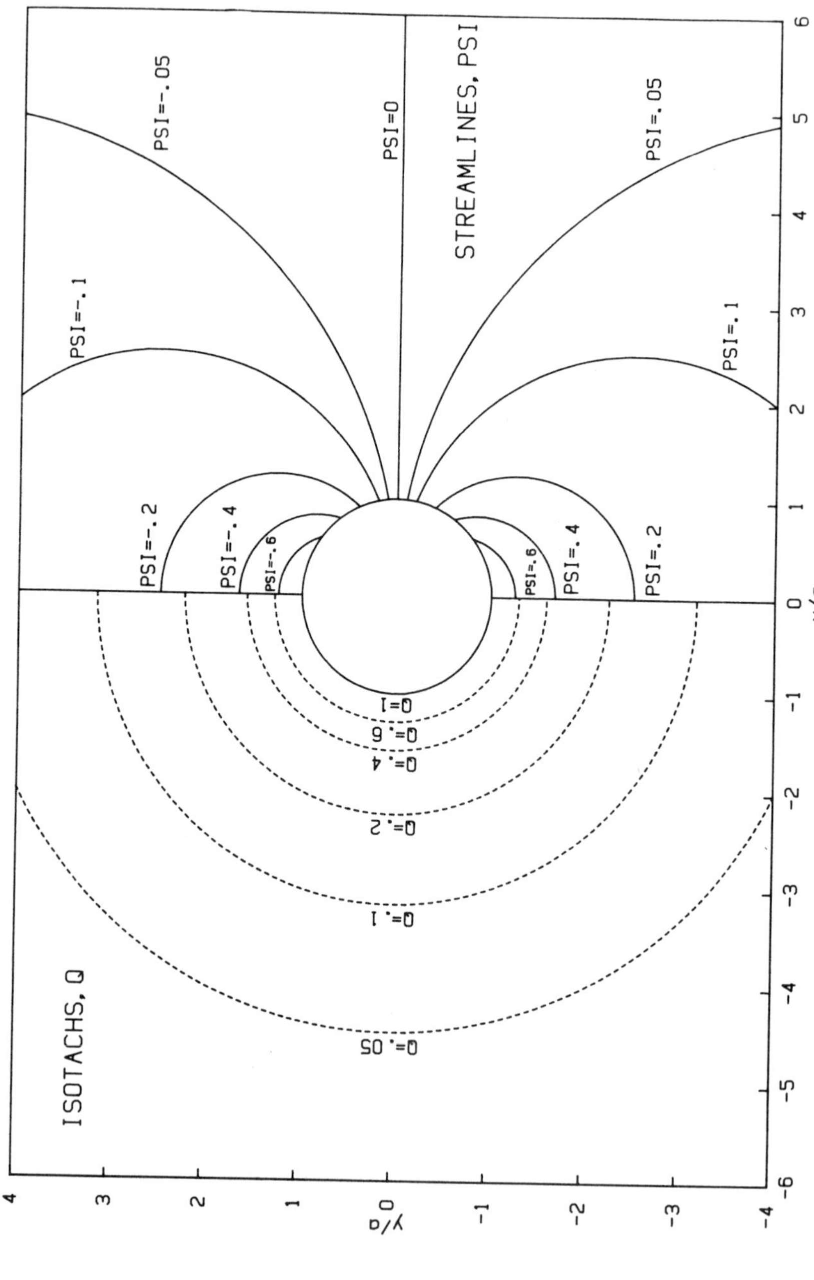

Figure 13 A translating cylinder.

Streamlines and isotachs are plotted at one instant of time on Figure 13. Note the similarity to Figure 6.

1.12 CYLINDER IN A TUNNEL

Milne-Thomson (1960), page 175, gives the complex potential for the flow of circulation $2\pi\kappa$ around a cylinder η_1 enclosed in a larger cylinder η_2 as

$$w = -\kappa\zeta \tag{33}$$

where $\zeta = \xi + i\eta$. The physical plane is related to the ζ plane through the coaxal coordinate transformation

$$z = ic \cot \frac{1}{2}\zeta \tag{34}$$

Streamlines are given by $\psi = -\kappa\eta$ and thus PSI = $-\psi/\kappa$. The radius of each cylinder is given by $r/c = 1/\sinh \eta$ and the center of each cylinder on the x axis is given by $x/c = \coth \eta$. Dimensionless streamlines are plotted on Figure 14 where the inner cylinder is $\eta_1 = 0.85$ and the outer cylinder is $\eta_2 = 0.35$. The natural symmetry of the problem has been used to draw only the upper half of each cylinder.

Formulating the complex velocity dw/dz from (33) and (34) and then the speed results in the following equation for the isotachs.

$$\frac{qc}{2\kappa} = \sin^2(\xi/2)\cosh^2(\eta/2) + \cos^2(\xi/2)\sinh^2(\eta/2) \tag{35}$$

The dimensionless speed was chosen to be $Q = qc/\kappa$. Isotachs are plotted on Figure 14 with the maximum speed on the inner cylinder of Q = 2.414 and the minimum speed on the inner cylinder of Q = 0.414.

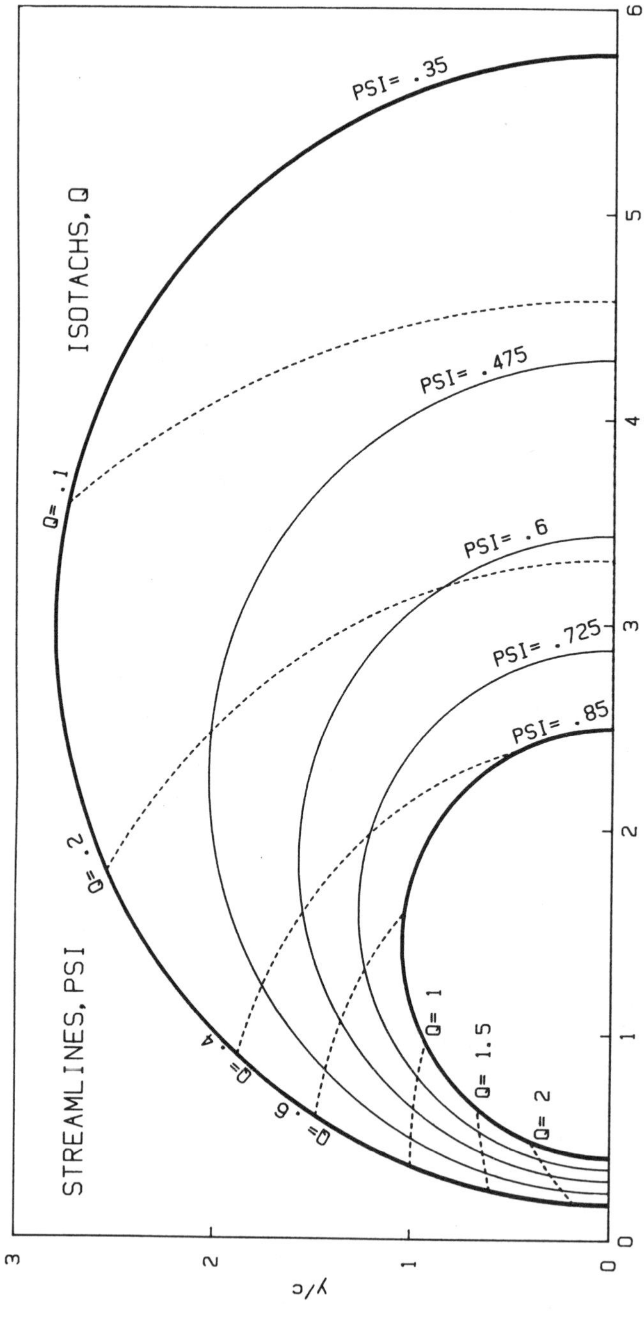

Figure 14 Circulation around a cylinder in a tunnel.

A natural extension of this problem is obtained by allowing the outer circle radius to approach infinity thus resulting in the circulation around a cylinder near a wall. Streamlines and isotachs for this flow are plotted in Figure 15 where now $\eta_1 = 1$ and $\eta_2 = 1 \times 10^{-7}$. The natural symmetry of the problem has again been utilized to plot only half of each cylinder.

1.13 SOURCE OUTSIDE A CYLINDER

Milne-Thomson (1960), page 213, gives the complex potential for a source of strength m located at z = f outside a circular cylinder of radius a centered at the origin.

$$w = -m \log (z - f) - m \log \left(\frac{a^2}{z} - f \right) \tag{36}$$

And in example 26, page 227, Milne-Thomson (1960) also gives the equation for the streamlines as

$$(x^2 + y^2)(cx + y) - (x^2 + y^2)c(f + f') + a^2(xc - y) = 0 \tag{37}$$

where $c = \tan (\psi/m)$, $f f' = a$. The dimensionless stream function is chosen PSI = ψ/m and all lengths are non dimensionalized by the cylinder radius. Equation (37) was solved numerically for y = f(x, PSI) using the Newton-Raphson technique (James, Smith and Wolford, 1967). Dimensionless streamlines for a source at x/a = 2 are plotted on the upper half of Figure 16. The effect of moving the source away from the cylinder is shown on Figure 17 where the upper half is a plot of the streamlines for a source at x/a = 10 and the lower half is for a source at x/a = 100.

Milne-Thomson (1960) also provides the equation for the speed

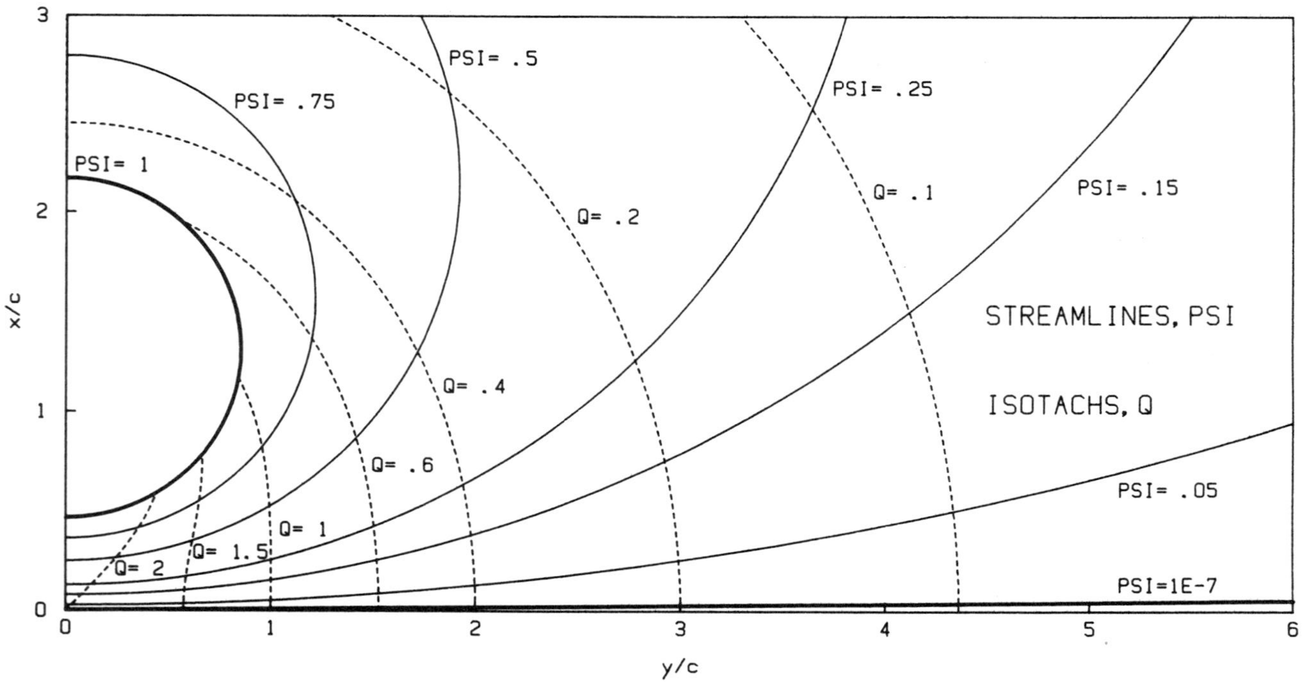

Figure 15 Circulation around a cylinder near a wall.

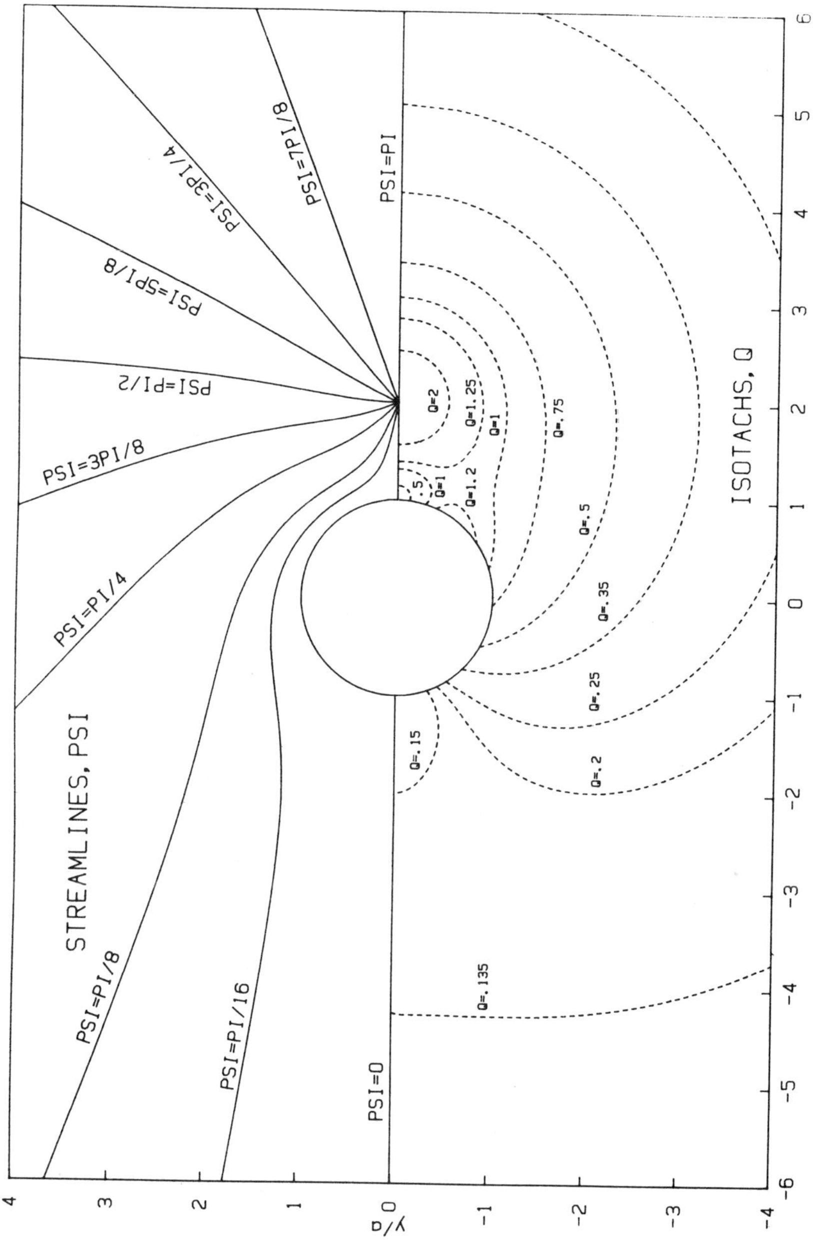

Figure 16 A cylinder and a source at $x/a = 2$.

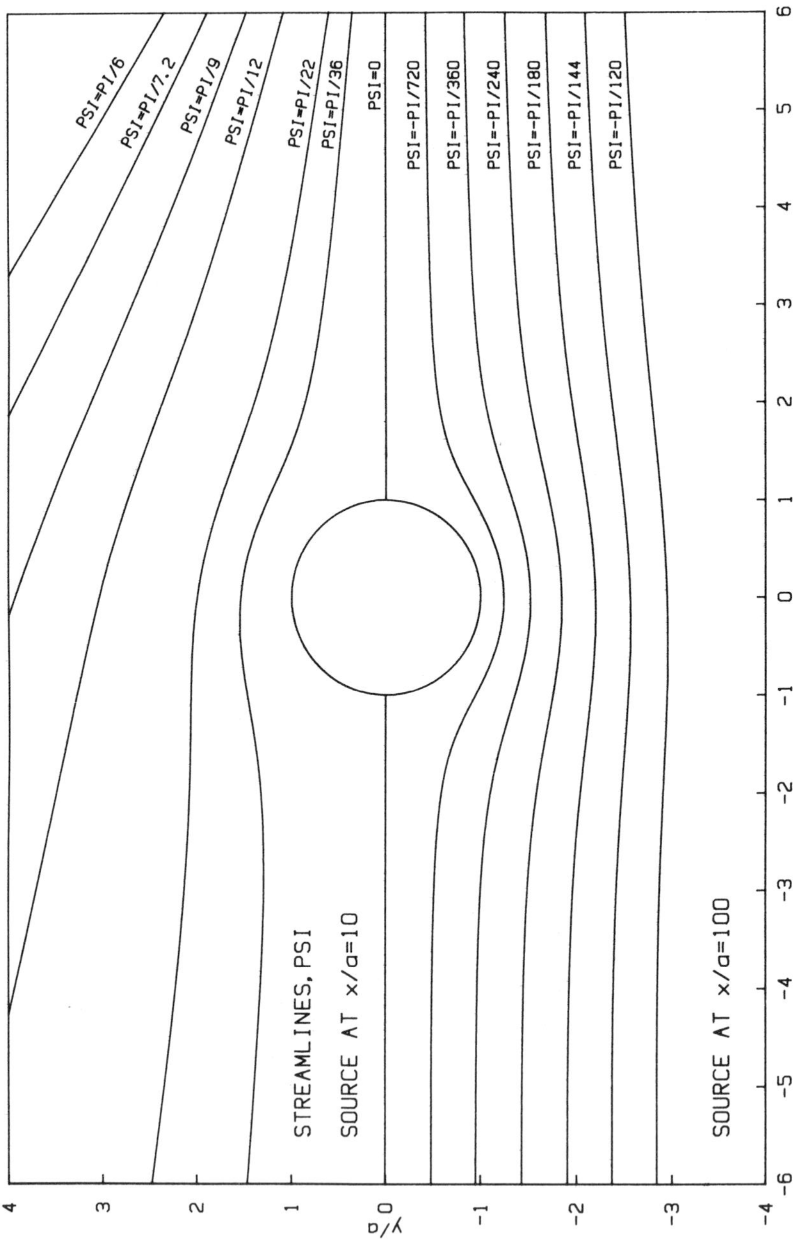

Figure 17 A cylinder and a source at x/a = 10, x/a = 100.

$$q = m \left| \frac{z^2 \mp a^2}{(z - f)z(z - a^2/f)} \right| \tag{38}$$

Equation (38) was solved numerically using the Newton-Raphson technique for $x = f(Q, y)$. The dimensionless isotachs $Q = qa/m$ are plotted on the lower half of Figure 16 for $x/a = 2$. The flow stagnates at $r = a$, $\theta = 0$ and then accelerates to a maximum of $Q = 1.3328$ at $\theta = \pm 36°$ on the cylinder and returns to stagnation at $\theta = \pi$. This flow pattern produces a force on the cylinder which attracts it to the source.

1.14 THE CYLINDER WITH CIRCULATION

Milne-Thomson (1960), page 181, gives the complex potential for uniform flow past a circular cylinder of radius a with circulation κ about the cylinder.

$$w = V\left(z + \frac{a^2}{z}\right) + i\,\kappa \log\left(\frac{z}{a}\right) \tag{39}$$

The magnitude of the circulation κ sets the stagnation points on the cylinder at angle β in the third and fourth quadrant as $\sin \beta = \kappa/2aV$. The dimensionless stream function is PSI = ψ/aV and the dimensionless streamline equation can be written from (39) as

$$\text{PSI} = \frac{y}{a}\left(1 - \frac{a^2}{r^2}\right) + 2 \sin \beta \log\left(\frac{r}{a}\right) \tag{40}$$

Equation (40) was solved numerically for $r = f(\theta, \text{PSI})$ and dimensionless streamlines are plotted for $\beta = 15°$ on the left half of Figure 18 and for $\beta = 90°$ on the left half of Figure 19.

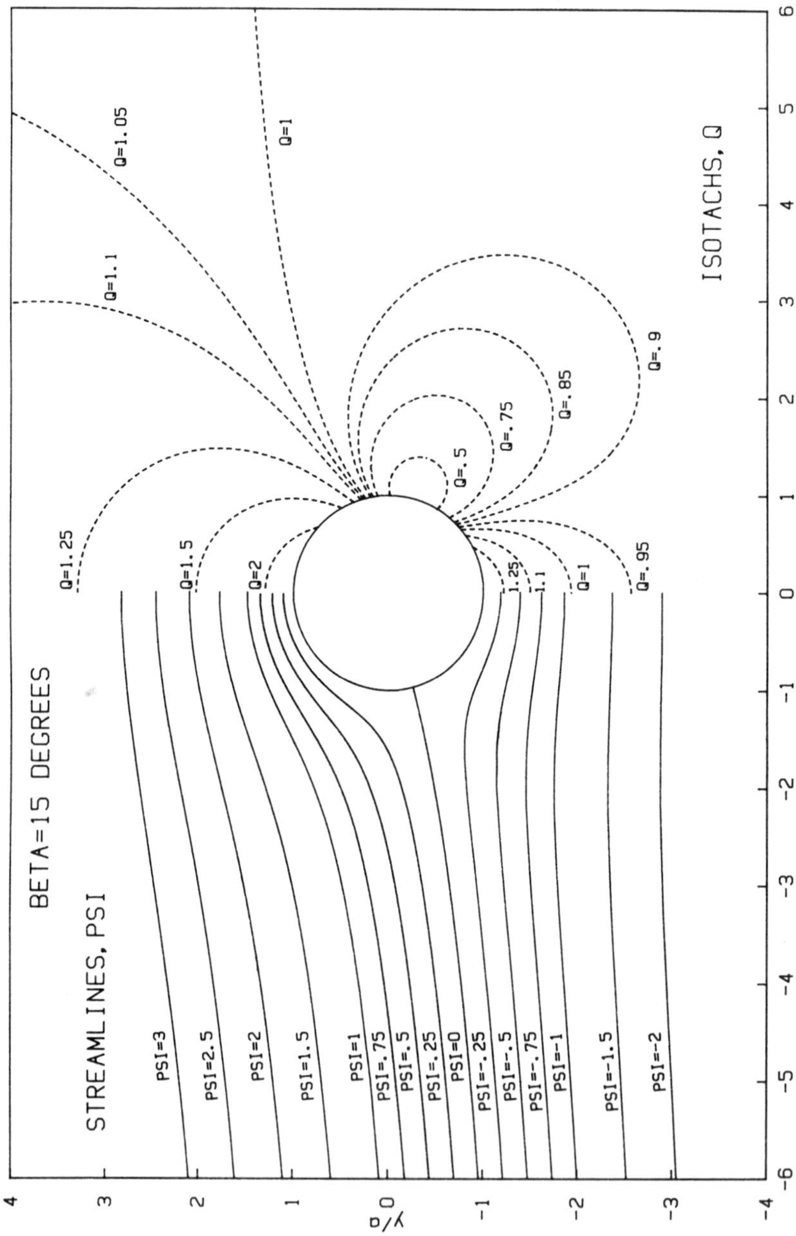

Figure 18 The cylinder with circulation, beta = 15°.

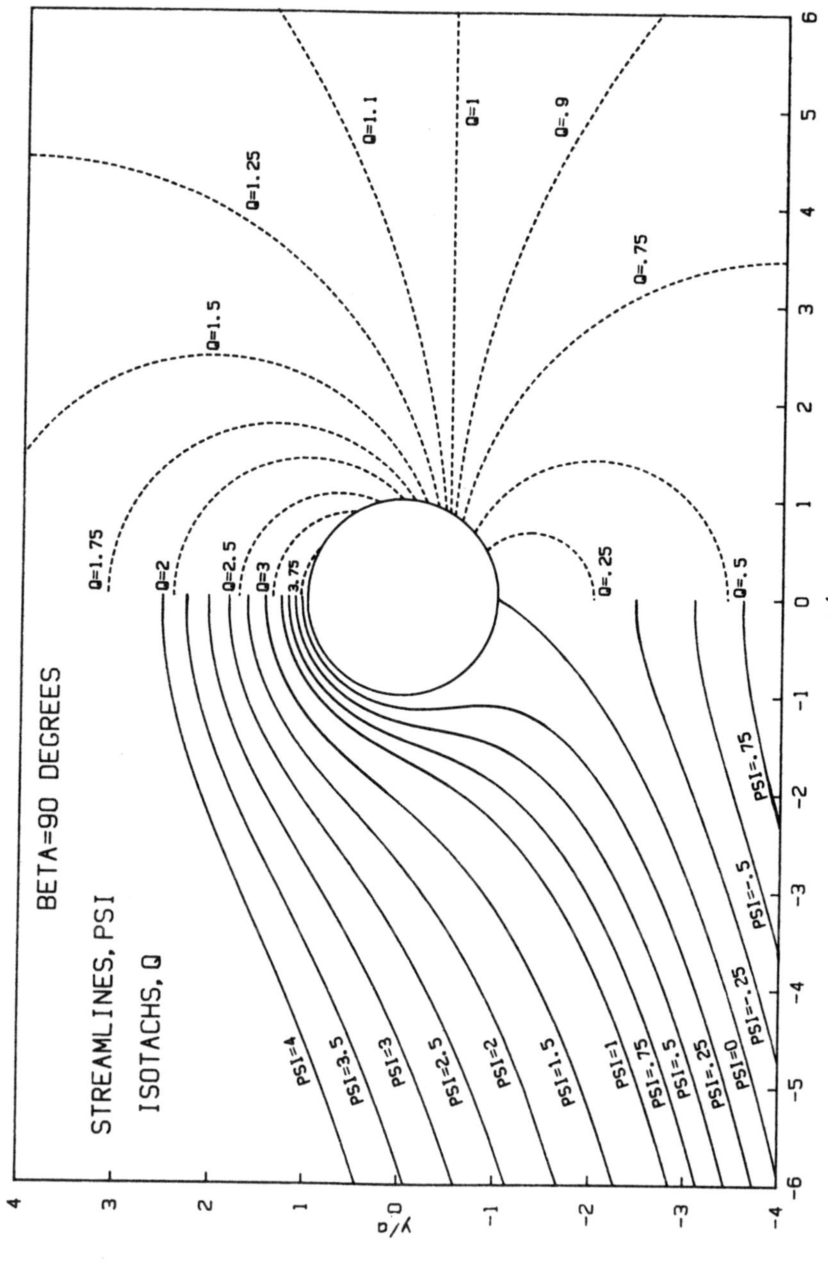

Figure 19 The cylinder with circulation, beta = 90°.

The complex velocity dw/dz and the speed q are developed from equation (39) as

$$Q = \left| \left(1 - \frac{a^2}{z^2}\right) + i\,\frac{2\,a\,\sin\,\beta}{z} \right| \tag{41}$$

where $Q = q/V$. Equation (41) was solved numerically for $r = f(Q, \theta)$ and isotachs for $\beta = 15°$ are plotted on the right half of Figure 18 and for $\beta = 90°$ on Figure 19.

The program listing for this problem is given in the Appendix where the details of the particular numerical solutions of (41) are given in subroutines 1580 and 1650.

1.15 THE CYLINDER IN A SHEAR FLOW

Milne-Thomson (1960), page 184, gives the stream function for a cylinder of radius a in a uniform shear flow of constant vorticity ω as

$$\psi = Vy - Va^2\,\frac{\sin\,\theta}{r} + \frac{1}{4}\,\omega r^2 \tag{42}$$

Equation (42) may be written in dimensionless format as

$$PSI = \frac{r}{a}\,\sin\,\theta\left(1 - \frac{a^2}{r^2}\right) + \frac{1}{2}\,\Omega\,\frac{r^2}{a^2} \tag{43}$$

where $PSI = \psi/Va$ and the dimensionless vorticity is $\Omega = \omega a/V$. From (43) it is seen that stagnation occurs on the cylinder at θ_S where

$$\sin\,\theta_S = -\frac{1}{4}\,\Omega \tag{44}$$

Equation (43) was solved numerically for $r = f(PSI, \theta)$ and streamlines are plotted on the right half of Figure 20 for the case of $\Omega = 0.1$ When interpreting Figure 20, observe that when $\Omega = 0.1$, the vorticity is sufficient to stagnate the flow at a distance of 10 cylinder radii below the cylinder. Also note that the stagnation streamline is from (43) $PSI = \Omega/4$.

The velocity components u_r and u_θ were calculated from (42) as

$$u_\theta = V \sin\left(1 + \frac{a^2}{r^2}\right) + \frac{1}{2}\omega r$$

$$u_r = -V \cos\left(1 - \frac{a^2}{r^2}\right) \tag{45}$$

The dimensionless speed $Q = q/V$ is formulated from (45) as

$$Q^2 = 1 + \left(\frac{a}{r}\right)^4 - 2\left(\frac{a}{r}\right)^2 \cos 2\theta + \Omega\left(\frac{r}{a}\right) \sin\theta\left(1 + \left(\frac{a}{r}\right)^2\right) + \left(\frac{\Omega r}{2a}\right)^2 \tag{46}$$

Isotachs are plotted on the left half of Figure 20. The maximum speed occurs on the top of the cylinder at a value $Q = 2 + \Omega/2$.

1.16 A DOUBLET AND ITS IMAGE

Milne-Thomson (1960), page 212, gives the complex potential for a doublet at $x = a$ and an equal antiparallel doublet at $x = -a$.

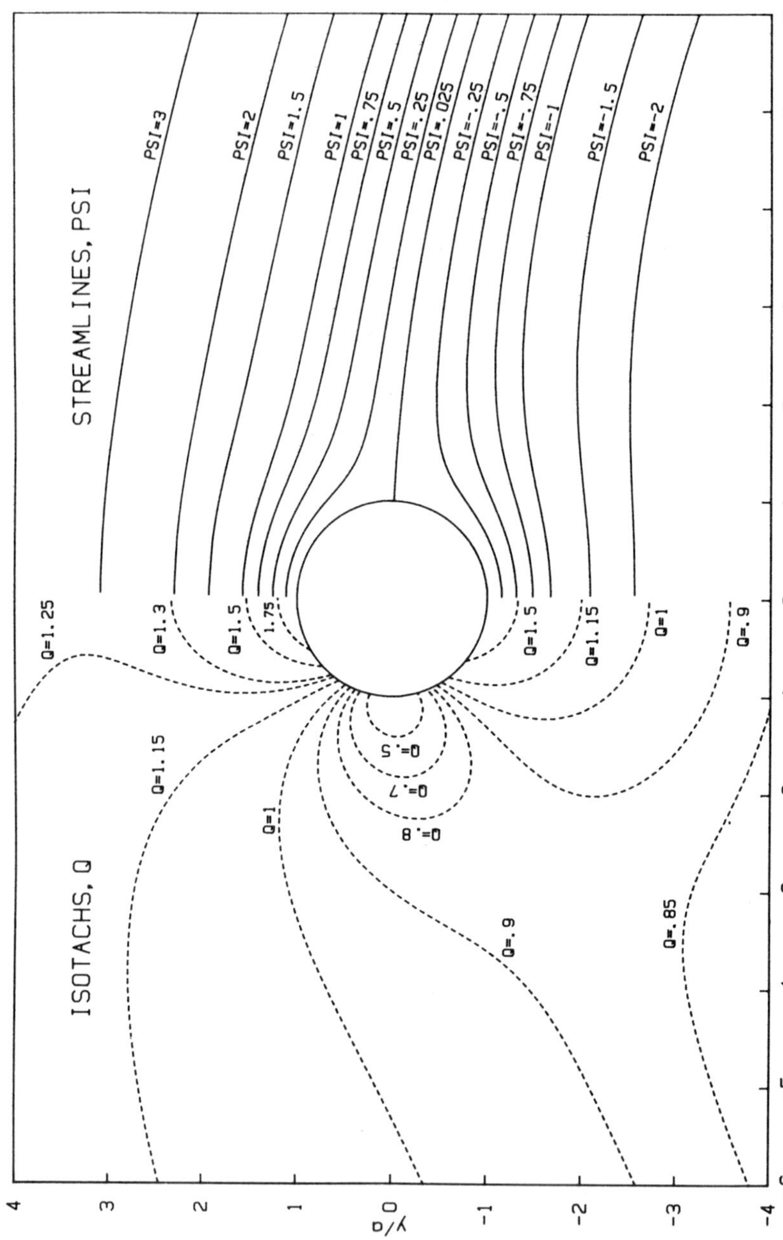

Figure 20 The cylinder in a shear flow.

$$w = \frac{\mu}{(z - a)} - \frac{\mu}{(z + a)} \tag{47}$$

The dimensionless stream function is PSI = $\psi a/\mu$ and the streamline equation may be written from (47) as

$$\text{PSI} = \frac{-2(r/a)^2 \sin 2\theta}{(r/a)^4 - 2(r/a)^2 \cos 2\theta + 1} \tag{48}$$

Equation (48) was solved using the quadratic formula for $r = f(\text{PSI}, \theta)$ and dimensionless streamlines are plotted on the right half of Figure 21. This is the flow pattern for a doublet near a wall.

Considerable simplification in the development of the isotachs is obtained by introducing elliptic coordinates through the transformation $z = a \cosh \zeta$ where $\zeta = \xi + i\eta$. Equation (47) becomes

$$w = \frac{\mu}{a \cosh \zeta - a} - \frac{\mu}{a \cosh \zeta + a} \tag{49}$$

When (49) is differentiated to formulate the complex velocity and the speed q the resulting isotach equation is

$$Q^2 = 16 \frac{|\cosh^2 \zeta|}{|\sinh^4 \zeta|} \tag{50}$$

where $Q = qa/\mu$. Equation (50) was solved for lines of constant Q by the Newton-Raphson technique and the results were transformed back to the physical plane through the parametric equations

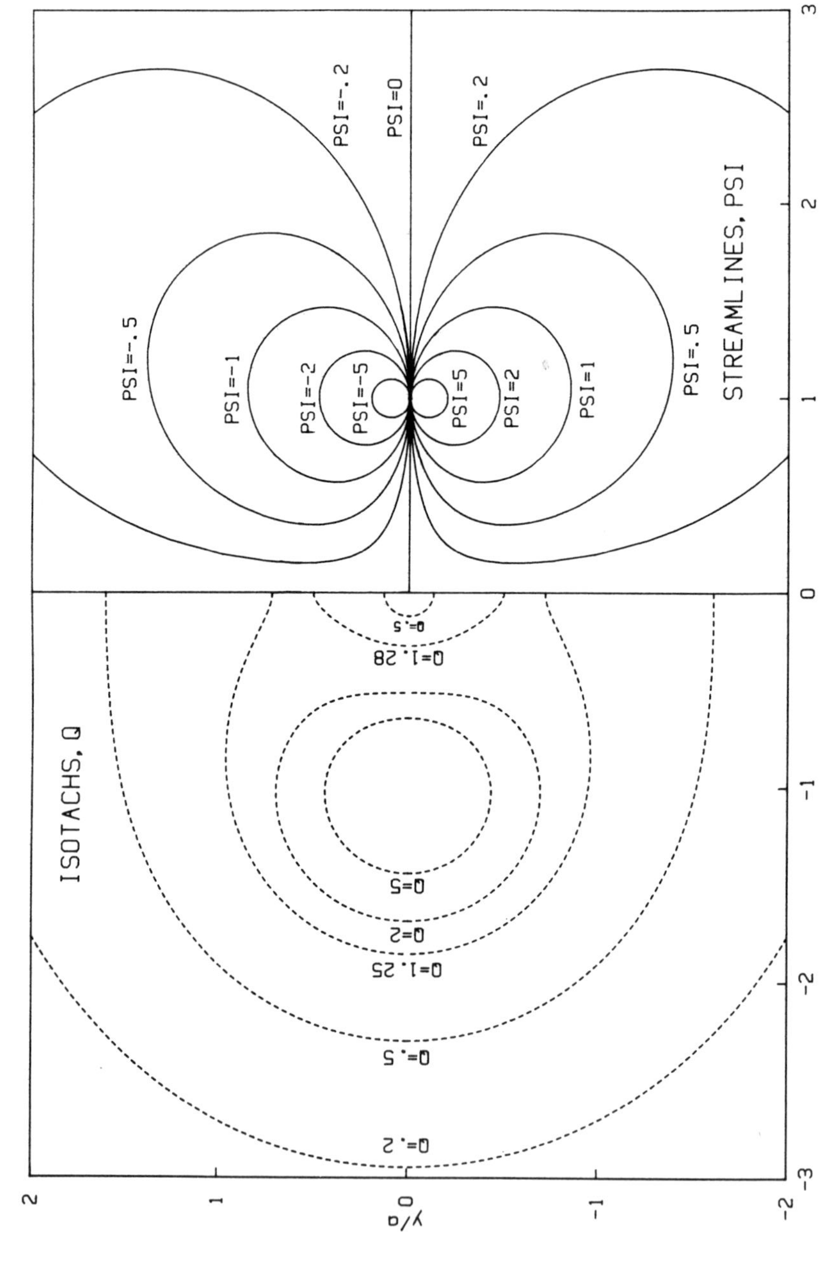

Figure 21 A doublet at x/a = 1 and its image.

$$x/a = \cosh\xi \cos\eta$$

$$y/a = \sinh\xi \sin\eta$$

Dimensionless isotachs are plotted on the left half of Figure 21 with a maximum speed on the wall of $Q = (3/4)(3)^{1/2}$ at $y/a = (1/3)^{1/2}$.

1.17 A ROW OF DOUBLETS IN A STREAM

Streeter (1948), page 116, gives the complex potential for an infinite row of doublets transverse to a uniform flow as

$$w = Uz + C' \coth(\pi z/a) \tag{51}$$

The doublets are spaced equidistant along the y axis with separation a. When the constant $C' = \pi b^2 U/a$ and a/b is sufficiently large, this potential represents the flow over an infinite row of cylinders of radius b separated by a distance a on the y axis. For this example, the value $b/a = 1/\pi$ was chosen. The dimensionless streamline equation is written from (51), with $PSI = \psi/Ub$, as

$$PSI = \frac{y}{b} - \frac{\pi b}{a} \frac{\sin(2\pi y/a)}{\cosh(2\pi x/a) - \cos(2\pi y/a)} \tag{52}$$

PSI = 0 in (52) generates the approximate cylindrical shape enclosing the doublets. For this example with $b/a = 1/\pi$, the body shape is almost a cylinder with a semidiameter on the x axis of $x/b = 0.8814$ and a semidiameter on the y axis of $y/b = 0.8603$. Lines of constant PSI were plotted from (52) on the upper half of Figure 22.

Equation (51) was differentiated to formulate the complex velocity and the speed as

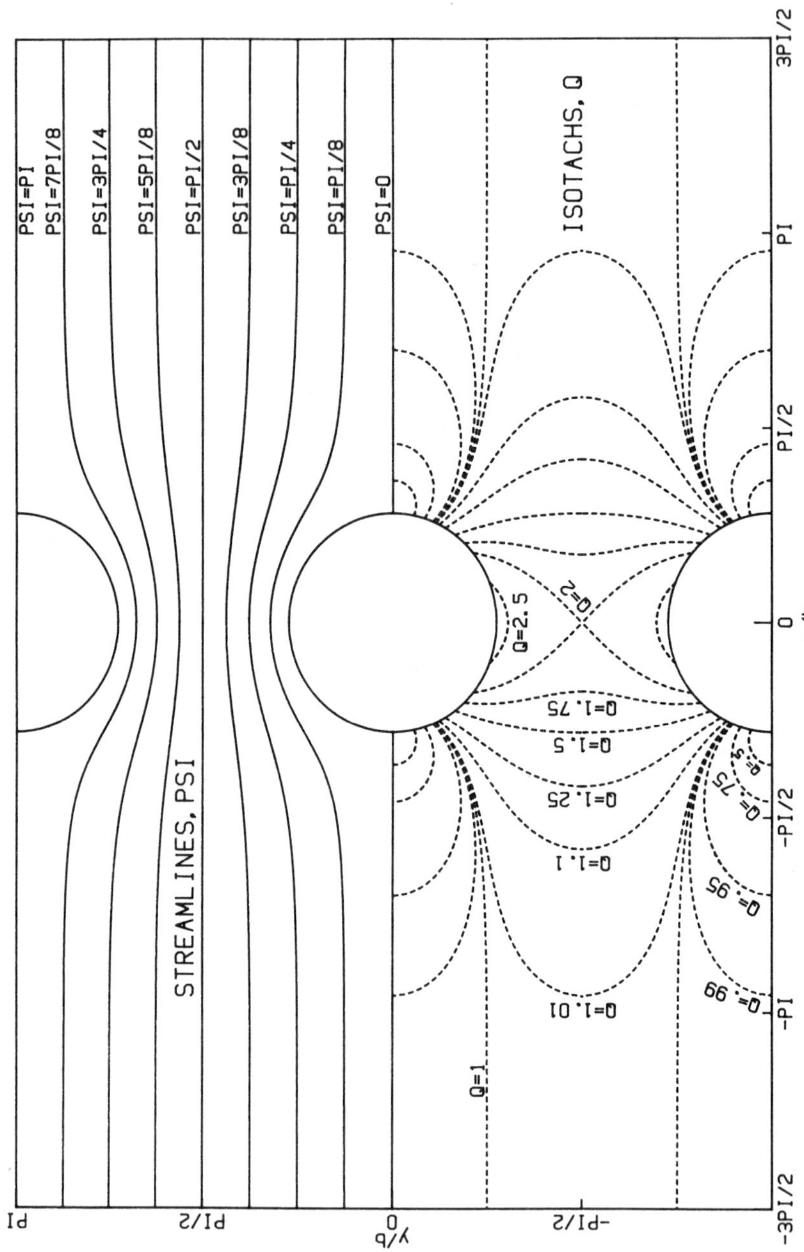

Figure 22 A row of doublets in a stream.

$$Q^2 = (1 - \operatorname{csch}^2(\pi z/a))(1 - \operatorname{csch}^2(\pi \bar{z}/a)) \tag{53}$$

where $Q = q/U$. Isotachs are plotted from (53), numerically solved using the Newton-Raphson technique, on the lower half of Figure 22. The maximum speed of the flow between the cylinders is $Q = 2.74$ at $x = 0$, $y = 0.8603$.

1.18 FLOW INTO A CANAL

Lamb (1932), page 74, gives the complex potential for the flow into a canal.

$$z = w + \exp(w) \tag{54}$$

Equation (54) can be modified slightly to scale the canal width to 2b with speed U at a point far into the canal.

$$\frac{\pi z}{b} = \frac{w\pi}{Ub} + \exp\left(\frac{w\pi}{Ub}\right) \tag{55}$$

The entrance plane of the canal is now $x = -b/\pi$. Dimensionless streamlines are lines of constant $PSI = \psi/Ub$ and (55) may be reduced to

$$\frac{x}{b} = \frac{1}{\pi} \ell n \left(\frac{\pi(y/b - PSI)}{\sin(\pi PSI)}\right) + \frac{y/b - PSI}{\tan(\pi PSI)}$$

Streamlines are plotted on the upper half of Figure 23.

Differentiating (55) to formulate the complex velocity and the speed yields

$$Q^{-2} = 1 + 2\exp(\pi PHI)\cos(\pi PSI) + \exp(2\pi PHI) \tag{56}$$

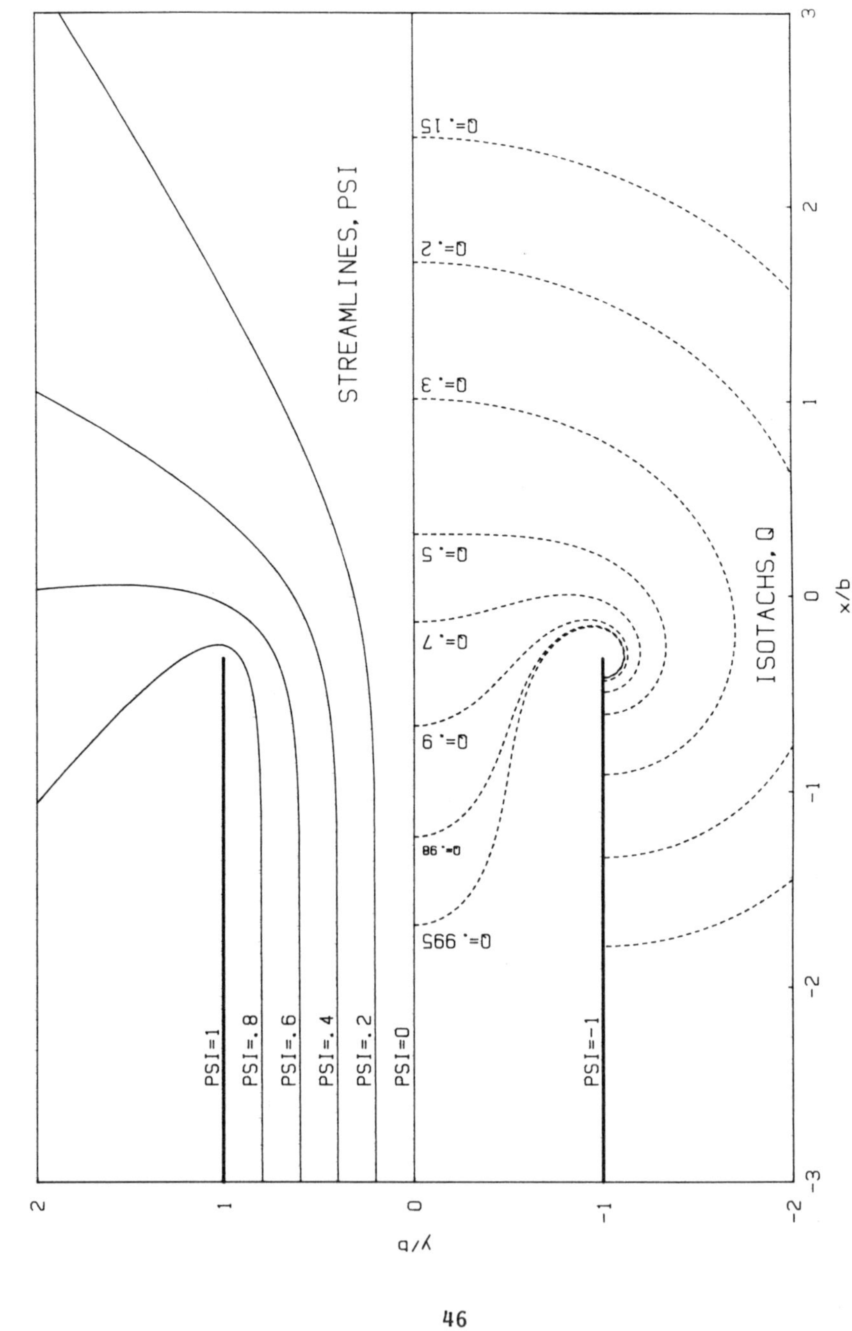

Figure 23 Flow into a canal of width 2b.

where Q = q/U and PHI = ϕ/Ub. Equation (56) was solved numerically using a Newton-Raphson technique and the isotachs are plotted on Figure 23. The speed approaches infinity at the sharp cornered entrance to the canal.

1.19 FLOW THROUGH AN APERTURE

Milne-Thomson (1960), page 152, gives the complex potential for flow through an aperture of width 2c which can be nondimensionalized to

$$\frac{z}{c} = \cosh \frac{w}{Uc} \tag{57}$$

The equation for the dimensionless streamlines PSI = ψ/U is written from (57) as

$$\frac{(x/c)^2}{\cos^2 \mathrm{PSI}} - \frac{(y/c)^2}{\sin^2 \mathrm{PSI}} = 1 \tag{58}$$

Streamlines plotted from (58) are seen to be hyperbolas as shown on the right half of Figure 24.

Differentiating (57) to formulate the complex velocity and the speed yields

$$Q^{-2} = \sinh(w/Uc)\ \sinh(\bar{w}/Uc) \tag{59}$$

where Q = q/U. Or, in terms of the physical variable z,

$$Q^{-4} = ((z/c)^2 - 1)((\bar{z}/c)^2 - 1) \tag{60}$$

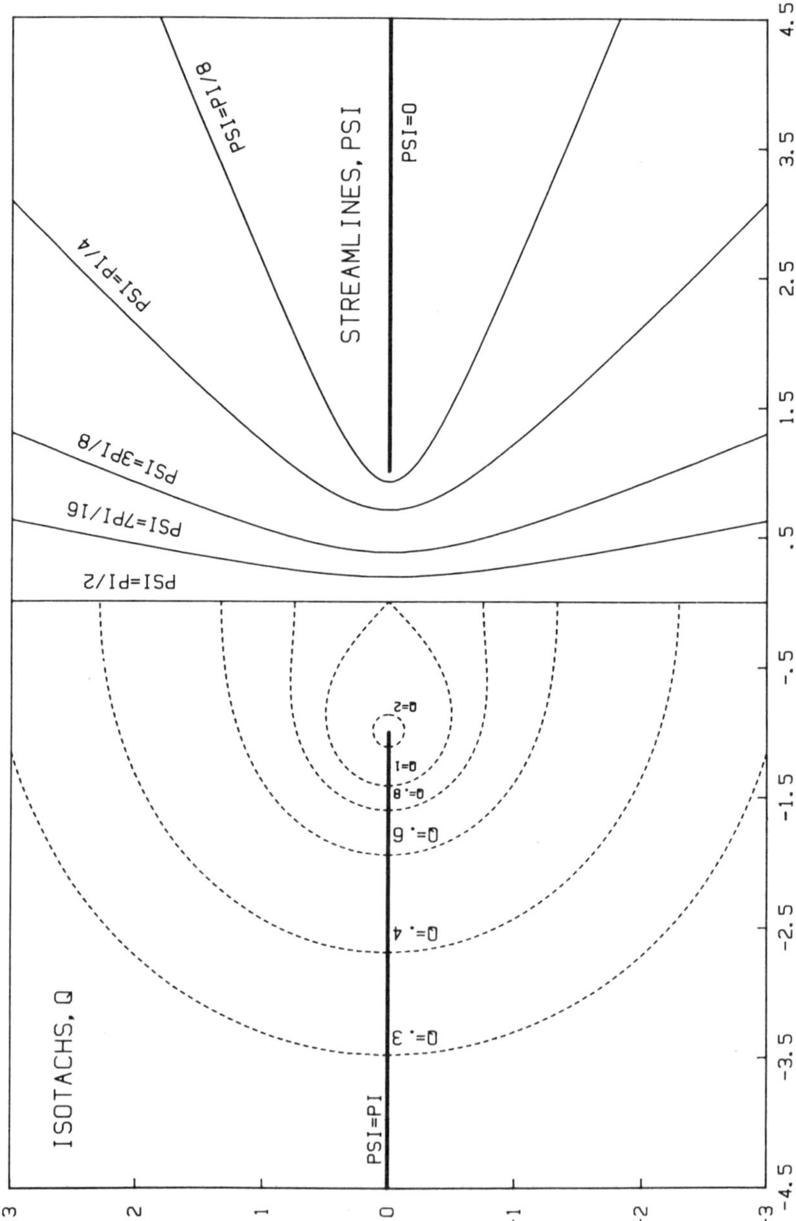

Figure 24 Flow through an aperture of width 2c.

The fourth order polynomial (60) was solved using the quadratic formula and isotachs are plotted on the left half of Figure 24. The speed approaches infinity at the sharp edged orifice but remains finite on any orifice formed by the streamlines $0 < PSI < \pi$.

1.20 FLOW OVER A LOG

Milne-Thomson (1960), page 174, gives the complex potential for the uniform flow U past a cylindrical log of radius a resting on the plane y = 0.

$$w = a\pi U \coth \frac{a\pi}{z} \tag{61}$$

The streamline equation can be written from (61), in terms of the dimensionless $PSI = \psi/Ua$, as

$$\frac{PSI}{\pi} = \frac{\cos A \sin A}{(\sinh B \cos A)^2 + (\cosh B \sin A)^2} \tag{62}$$

where $A = (\pi a/r) \sin \theta$ and $B = (\pi a/r) \cos \theta$.

Equation (62) was solved numerically for $r = f(\theta, PSI)$. Dimensionless streamlines are plotted on the left half of Figure 25 where PSI = 0 is both the log and the plane y = 0.

The equation for the isotachs is also given as

$$Q^2 = \left(\frac{a\pi}{r}\right)^4 \frac{2}{\left(\cosh\left(\frac{2a\pi x}{r^2}\right) - \cos\left(\frac{2a\pi y}{r^2}\right)\right)} \tag{63}$$

where $Q = q/U$. Isotachs are plotted on the right half of Figure 25. The flow stagnates at the bottom of the log then accelerates through Q = 1 at $\theta = 49.2°$ and then to a maximum $Q = (\pi/2)^2$ on the top of the log.

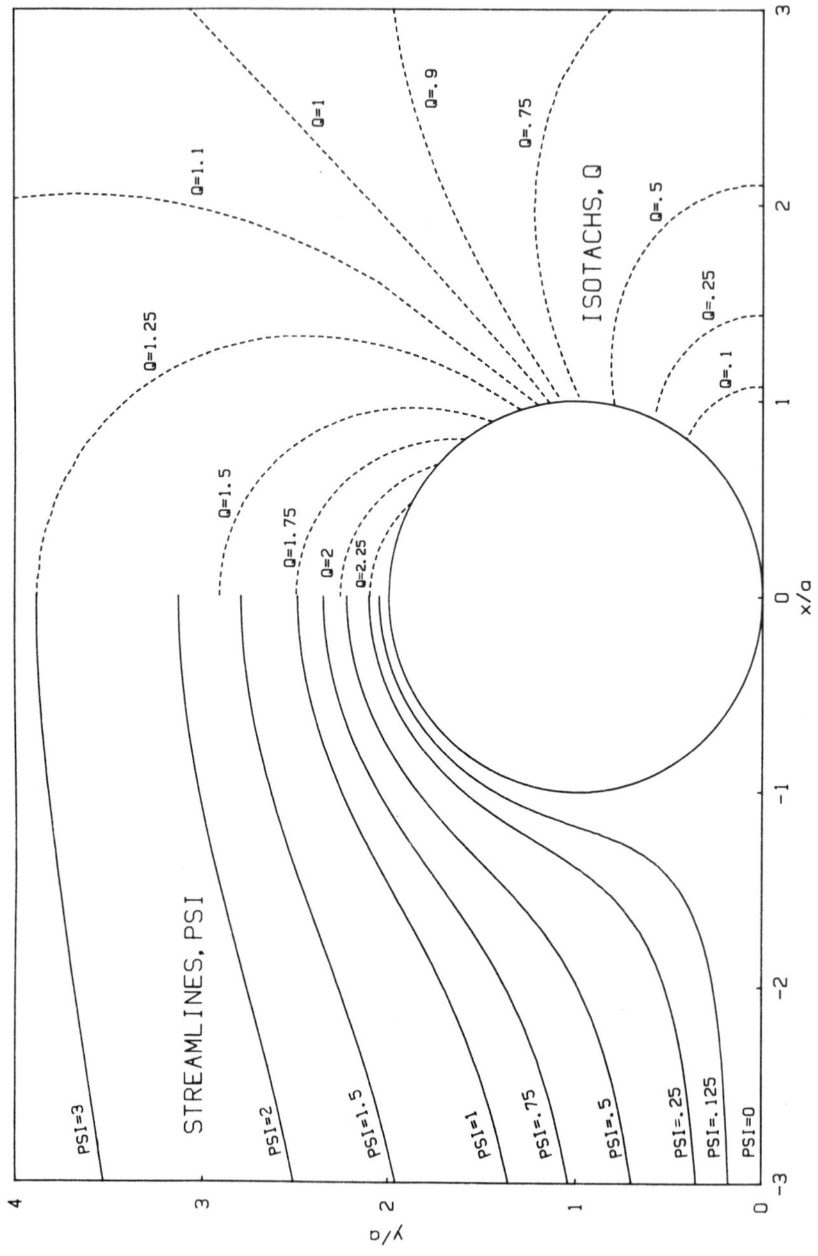

Figure 25 Flow over a log.

1.21 DITCH AND MOUND FLOW

Milne-Thomson (1960), page 172, gives the complex potential

$$w = U \frac{2ci}{n} \cot \frac{\zeta}{n} \tag{64}$$

which is linked to the physical plane through the coaxal coordinate transformation $z = ic \cot(\zeta/2)$ where $\zeta = \xi + i\eta$. When $n = 1.5$, (64) is the potential for the flow over a mound of width 2c and maximum height $y/c = 0.4142$. When $n = 2.5$, (64) is the potential for flow over a shallow ditch 2c wide and maximum depth $y/c = 0.4142$. And finally, when $n = 3$, (64) is the potential for flow over a semicircular ditch of radius $r/c = 1$. The dimensionless streamlines are defined by $PSI = \psi/Uc$ and the streamline equation is developed from (64) as

$$\frac{n\,PSI}{2} = \frac{\sin(\xi/n)\cos(\xi/n)}{(\sin(\xi/n)\cosh(\eta/n))^2 + (\cos(\xi/n)\sinh(\eta/n))^2} \tag{65}$$

Equation (65) was solved directly for $\eta = f(n, \xi, PSI)$ to generate the streamlines. This solution was linked to the physical plane through the parametric equations

$$\frac{x}{c} = \frac{\sinh \eta}{\sin \eta - \cos \xi}; \quad \frac{y}{c} = \frac{\sin \xi}{\cosh \eta - \cos \xi} \tag{66}$$

Streamlines for flow over a mound, a shallow ditch and a semicircular ditch are shown on the left half of Figures 26, 27, and 28, respectively.

The equation for the isotachs is also given as

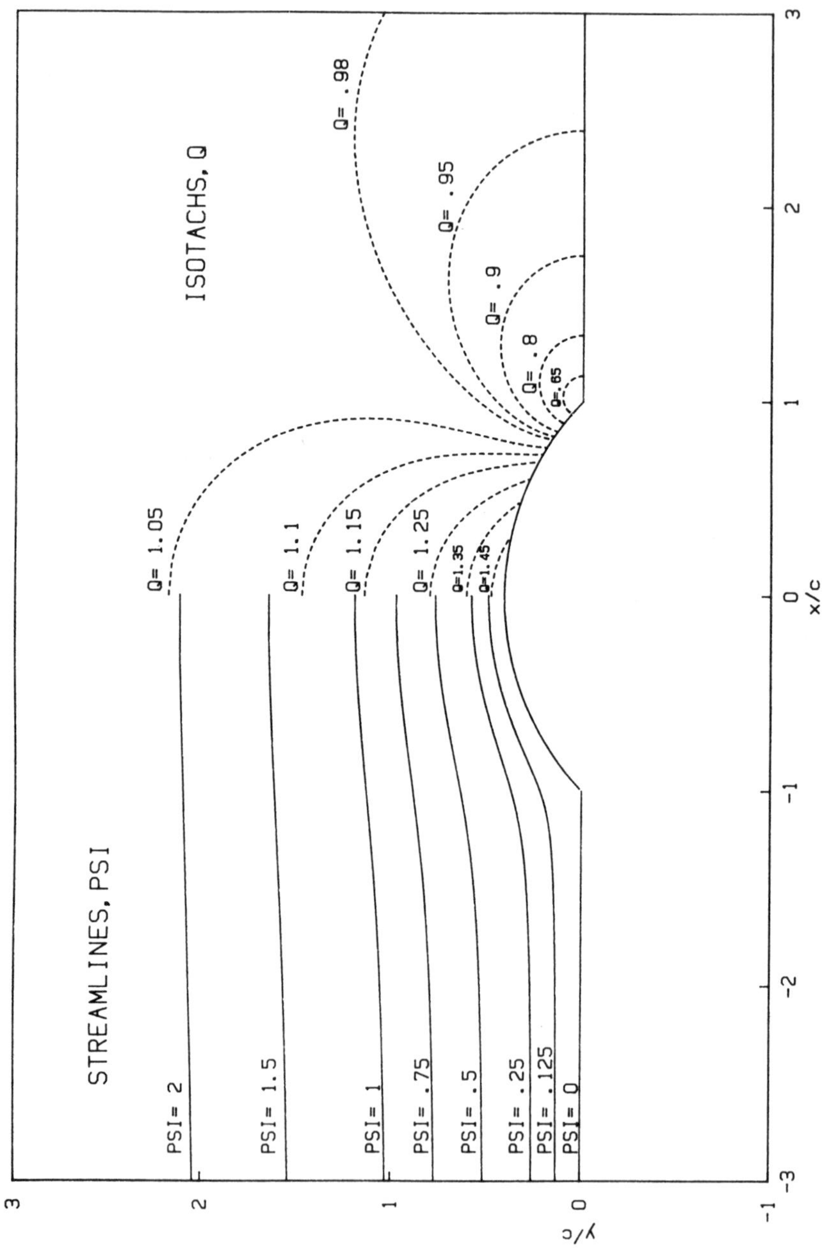

Figure 26 Flow over a mound.

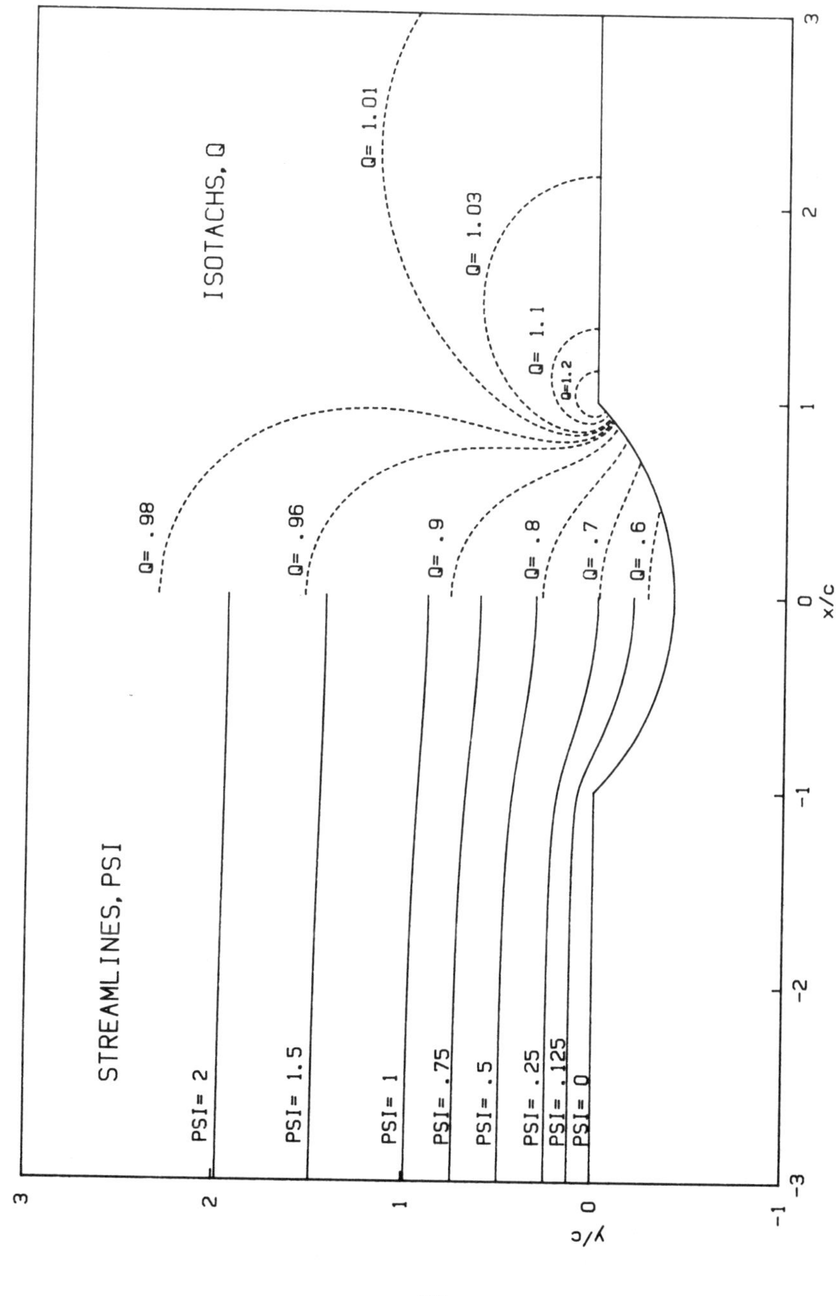

Figure 27 Flow over a shallow ditch.

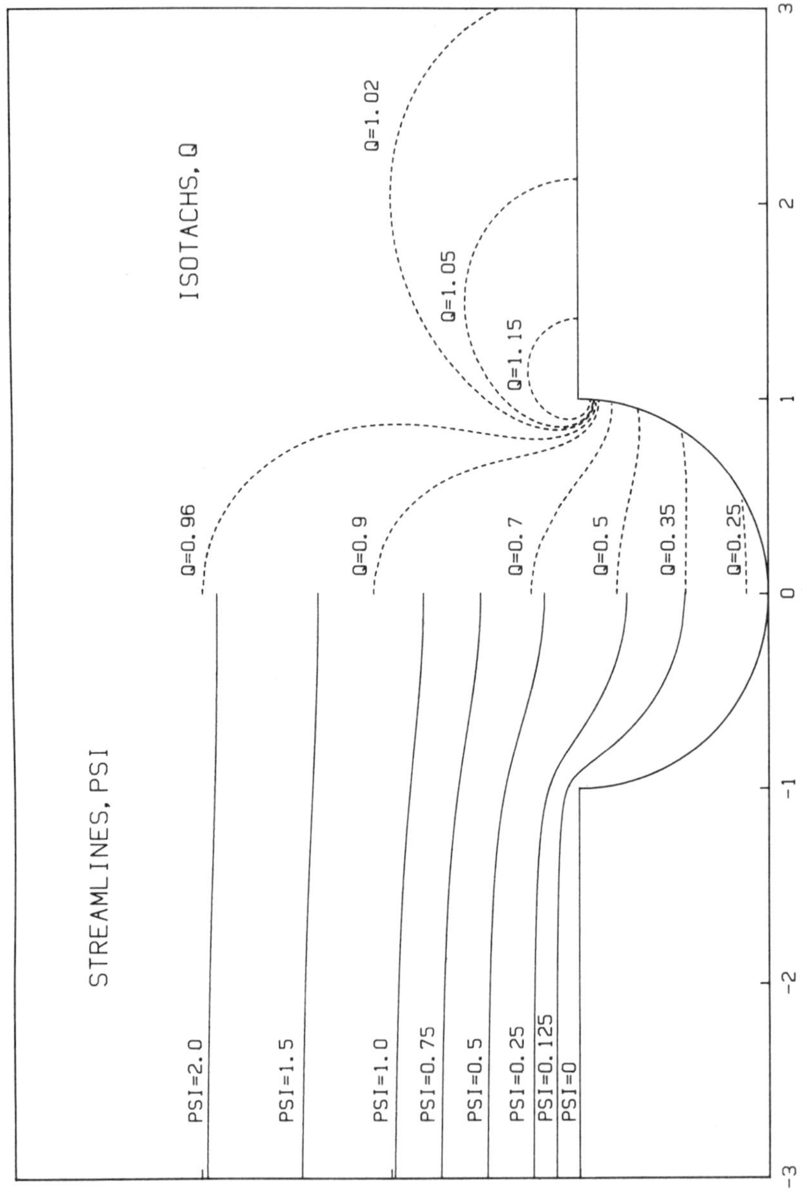

Figure 28 Flow over a semi-circular ditch.

$$Q^2 = \frac{16}{n^4}\left(\frac{\cosh \eta - \cos \xi}{\cosh(2\eta/n) - \cos(2\xi/n)}\right)^2 \tag{67}$$

where $Q = q/U$.

Equation (67) was solved numerically using a Newton-Raphson technique for $\eta = f(n, \xi, Q)$ and the results were transformed into the physical plane using the parametric equations (66). Isotachs for flow over a mount ($n = 1.5$) are plotted on the right half of Figure 26. The flow stagnates at $x = \pm c$, $y = 0$ and reaches a maximum speed on top of the mound of $Q = 1.51$. Isotachs for flow over a shallow ditch ($n = 2.5$) are plotted on the right half of Figure 27. The minimum speed is $Q = 0.5463$ at the bottom of the ditch and the speed approaches infinity at the corners $x = \pm c$, $y = 0$. Isotachs for flow over a semicircular ditch ($n = 3$) are plotted on the right half of Figure 28. The minimum speed is $Q = 0.2222$ at the bottom of the ditch and the speed approaches infinity at the corners $x = \pm c$, $y = 0$.

1.22 FLOW OVER AN ELLIPSE

Milne-Thomson (1960), page 164, gives the complex potential for flow over an ellipse

$$w = U(a + b) \cosh(\zeta - \xi_o - i\alpha) \tag{68}$$

where α is the angle of attack, and a,b are the semi-axes major and minor, respectively. The elliptic coordinates are related to the physical plane through the transformation $z = c \cosh \zeta$ where $\zeta = \xi + i\eta$ and 2c is the distance between the focii. When $\xi = \xi_o$, the point x, y lies on the ellipse. The example plotted here is for $\alpha = 0$, $c = 1$, $a = 1.3416$, and $b = 0.8944$. The dimensionless streamlines were chosen to be $PSI = \psi/(a + b)U$ and the stream function can be written from (68)

$$\text{PSI} = \sinh(\xi - \xi_o) \sin \eta \tag{69}$$

Equation (69) was solved directly for $\xi = f(\eta, \text{PSI})$ and then transformed to the physical plane through the parametric equations $x/c = \cosh \xi \cos \eta$ and $y/c = \sinh \xi \sin \eta$. Streamlines are plotted on the upper half of Figure 29.

The isotach equation is also given as

$$Q^2 = \frac{(a + b)}{(a - b)} \frac{\sinh^2(\xi - \xi_o) + \sin^2 \eta}{\sinh^2 \xi + \sin^2 \eta} \tag{70}$$

where $Q = q/U$. Equation (70) was solved directly for $\eta = f(\xi, Q)$ and then transformed to the physical plane to produce the isotachs drawn on the lower half of Figure 29. The flow stagnates at the nose of the ellipse and then accelerates through $Q = 1$ to a maximum of $Q = 1.6667$ on the top.

1.23 FLOW PAST A FLAT PLATE; ALPHA = 45°

Milne-Thomson (1960), page 166, gives the complex potential for the flow past a flat plate of length 2a situated at angle α to a uniform flow U.

$$w = Ua \cosh(\zeta - i\alpha) \tag{71}$$

The flow is related to the physical plane through the elliptic coordinate transformation $z = a \cosh \zeta$. The stream function equation is written directly from (71) as

$$\text{PSI} = \sinh \xi \sin (\eta - \alpha) \tag{72}$$

where $\text{PSI} = \psi/Ua$. Equation (72) was solved directly for $\xi = f(\text{PSI}, \eta, \alpha)$ and the results were transformed to the physical plane through $x/a = \cosh \xi \cos \eta$ and $y/a = \sinh \xi \sin \eta$.

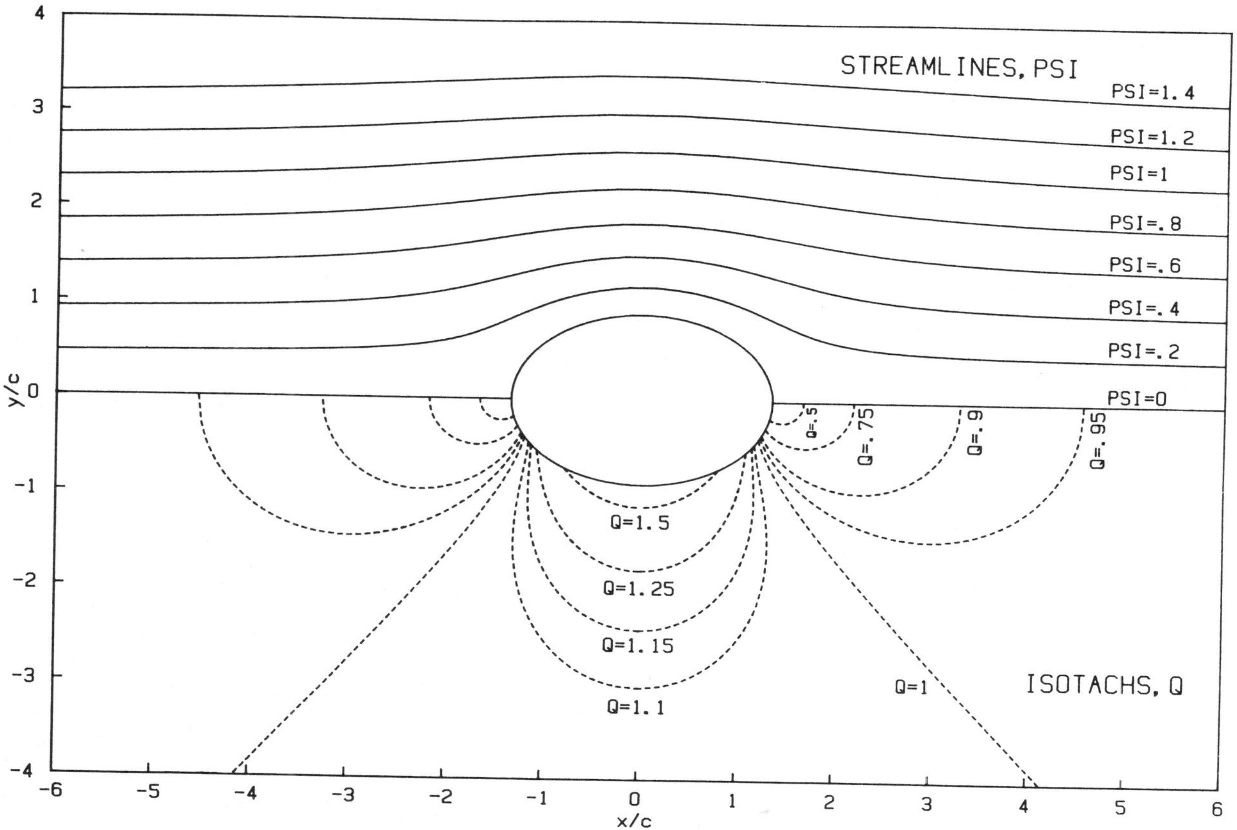

Figure 29 Flow over an ellipse.

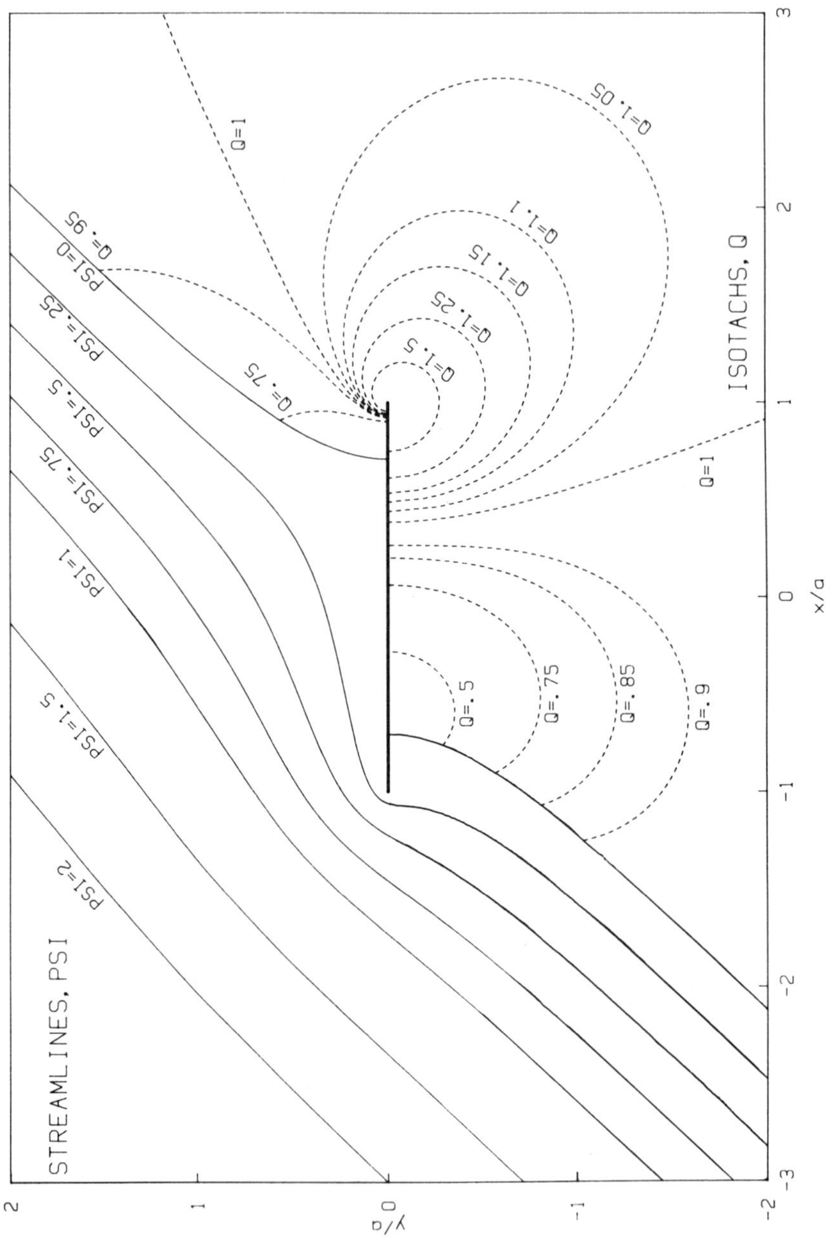

Figure 30 Flow over a flat plate, alpha = 45°.

For the case of $\alpha = \pi/4$, streamlines are plotted on Figure 30. The natural symmetry of the problem has been utilized to separate streamlines and isotachs along the line PSI = 0.

Formulating the complex velocity and the speed from (71) gives

$$Q^2 = \frac{\sinh^2\xi + \sin^2(\eta - \alpha)}{\sinh^2\xi + \sin^2\eta} \tag{73}$$

where $Q = q/U$.

Equation (73) was solved directly for $\xi = f(Q, \eta, \alpha)$, and the results were transformed to the physical plane and plotted on Figure 30. This flow has two stagnation points on the plate and the speed approaches infinity at both sharp corners.

1.24 A VORTEX FILAMENT

Milne-Thomson (1960), page 354, gives the complex potential for a vortex filament of strength κ at the origin.

$$w = i\,\kappa \log z \tag{74}$$

The strength κ is positive when circulation around the filament is counter-clockwise. The dimensionless stream function is $\text{PSI} = \psi/\kappa$ and the equation for the streamlines is written from (74) as

$$\text{PSI} = \cos(\theta)\log(r/a)$$

Dimensionless streamlines are circles centered at the origin and plotted on the upper half of Figure 31.

The complex velocity and the speed are developed from (74) as

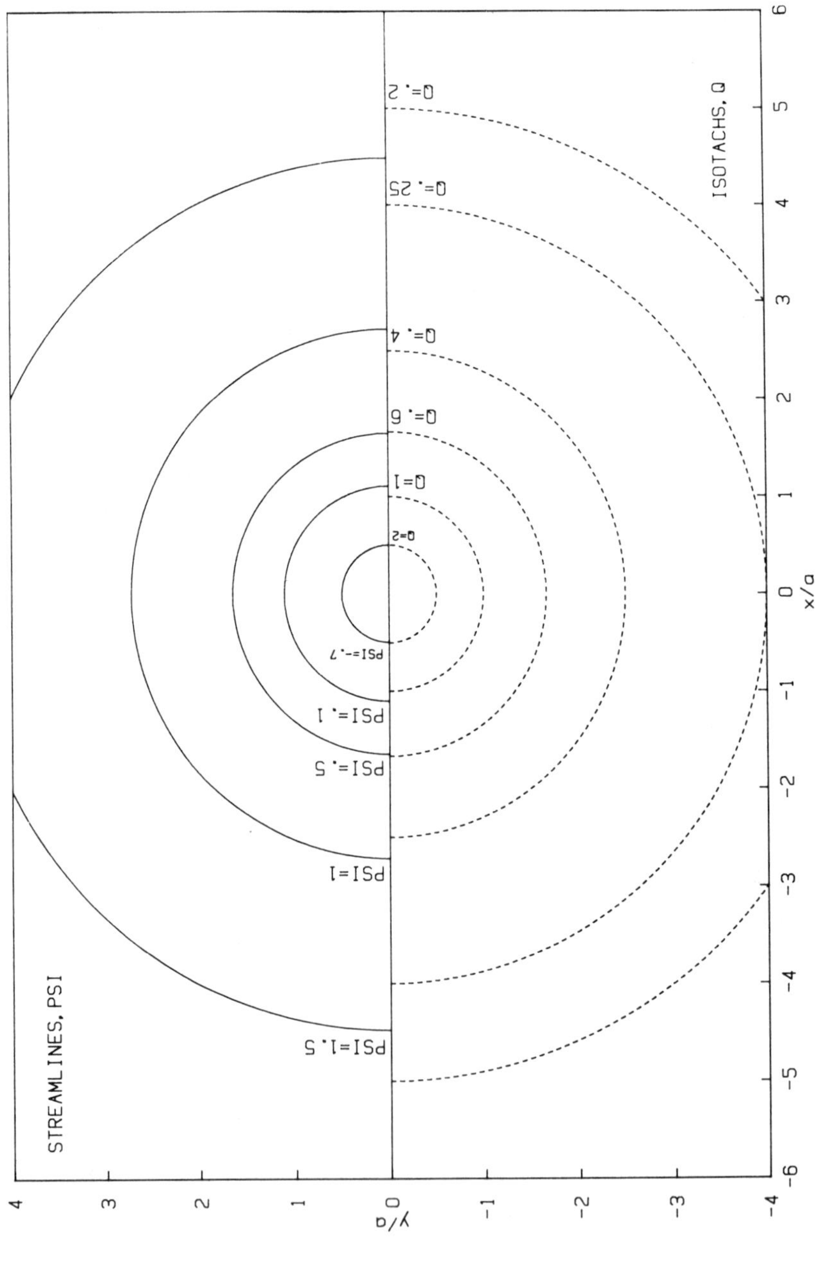

Figure 31 A vortex filament at the origin.

$$Q = a/r$$

where $Q = qa/\kappa$. Isotachs are also circles centered at the origin as plotted on the lower half of Figure 31.

1.25 A VORTEX PAIR OF OPPOSITE ROTATION

Milne-Thomson (1960), page 357, gives the complex potential for a vortex pair each of strength κ but of opposite rotation. The counter-clockwise vortex is located at 0, ia and the clockwise vortex is located at 0,-ia.

$$w = i\,\kappa \log((z - ia)/(z + ia)) \tag{75}$$

The dimensionless stream function is defined by $PSI = \psi/\kappa$ and the streamline equation is written from (75) as

$$PSI = \frac{1}{2}\log\left(\frac{x^2 + (y - a)^2}{x^2 + (y + a)^2}\right) \tag{76}$$

Streamlines are coaxal circles as shown on the upper half of Figure 32 for the counterclockwise vortex.

The complex velocity and the speed are formulated from (75) as

$$Q^2 = \frac{4}{((z/a)^2 + 1)((\bar{z}/a)^2 + 1)} \tag{77}$$

The isotach equation (77) is a fourth order polynomial which was solved using the quadratic formula. Isotachs are plotted on the lower half of Figure 32. This vortex pair experiences a translation in the positive x direction of speed $Q = 1/2$; thus, Figure 32 is a plot at one instant of time.

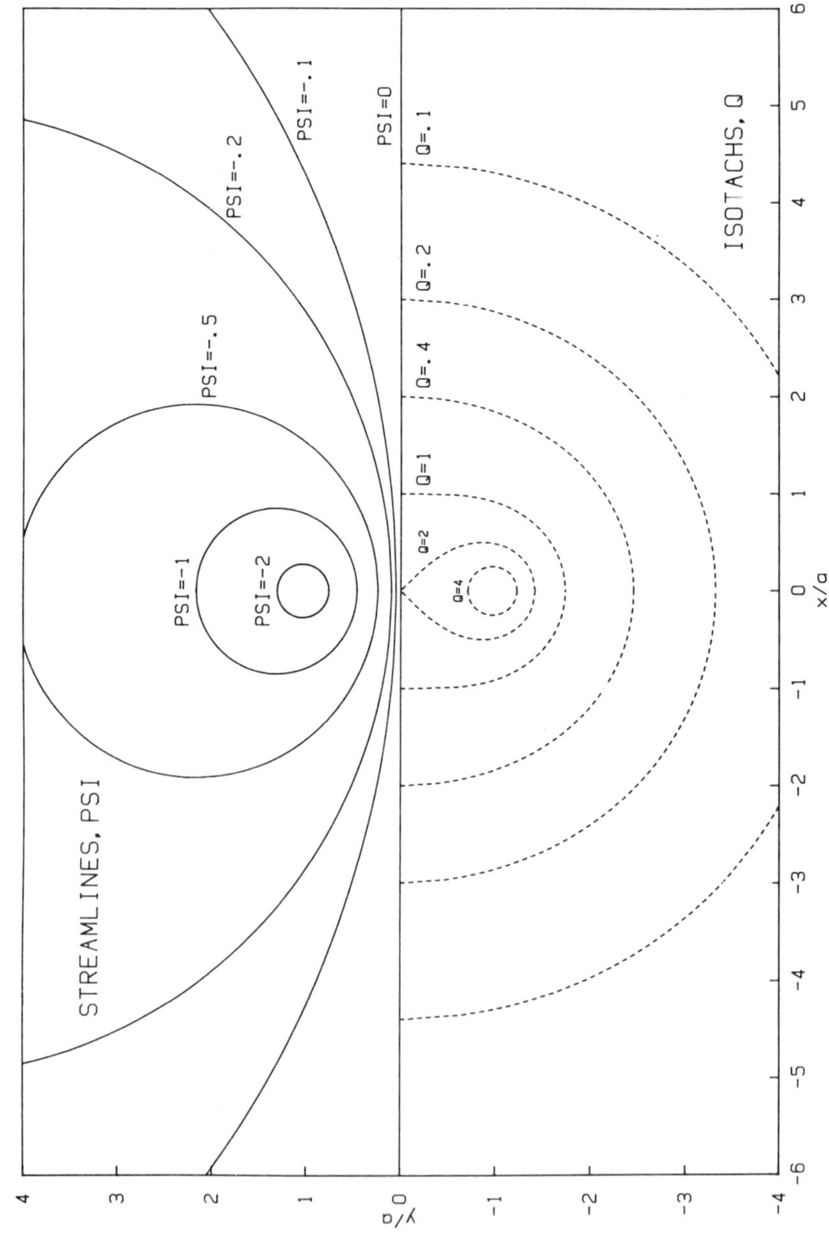

Figure 32 A vortex pair of opposite rotation.

1.26 A VORTEX PAIR OF IDENTICAL ROTATION

Milne-Thomson (1960), page 356, gives the complex potential for a vortex pair of identical rotation. At the instant that one vortex is at 0,ia and the other vortex is a 0,-ia, the complex potential may be written

$$w = i\,\kappa \log\left((z - ia)(z + ia)\right) \tag{78}$$

The dimensionless stream function is PSI = ψ/κ and the streamline equation is written from (78) as

$$\mathrm{PSI} = \log(r_1 r_2) \tag{79}$$

where $r_1 = (x^2 + (y - a)^2)^{1/2}$ and $r_2 = (x^2 + (y + a)^2)^{1/2}$. Equation (79) is a fourth order polynomial in x, y of the same form as (76) and it was solved using the quadratic formula. Streamlines are shown on the upper half of Figure 33. Their shape is identical to the isotachs of Figure 32.

Differentiating (78) to formulate the complex velocity and the speed yields

$$Q^2 = \frac{z\,\bar{z}/a^2}{((z/a)^2 + 1)((\bar{z}/a)^2 + 1)} \tag{80}$$

Equation (80) was solved for the isotachs using the quadratic formula as plotted on the lower half of Figure 33. This vortex pair rotates about the origin where there is a stagnation point. The tangential speed at each vortex is Q = 2 and Figure 33 is plotted at an instant of time.

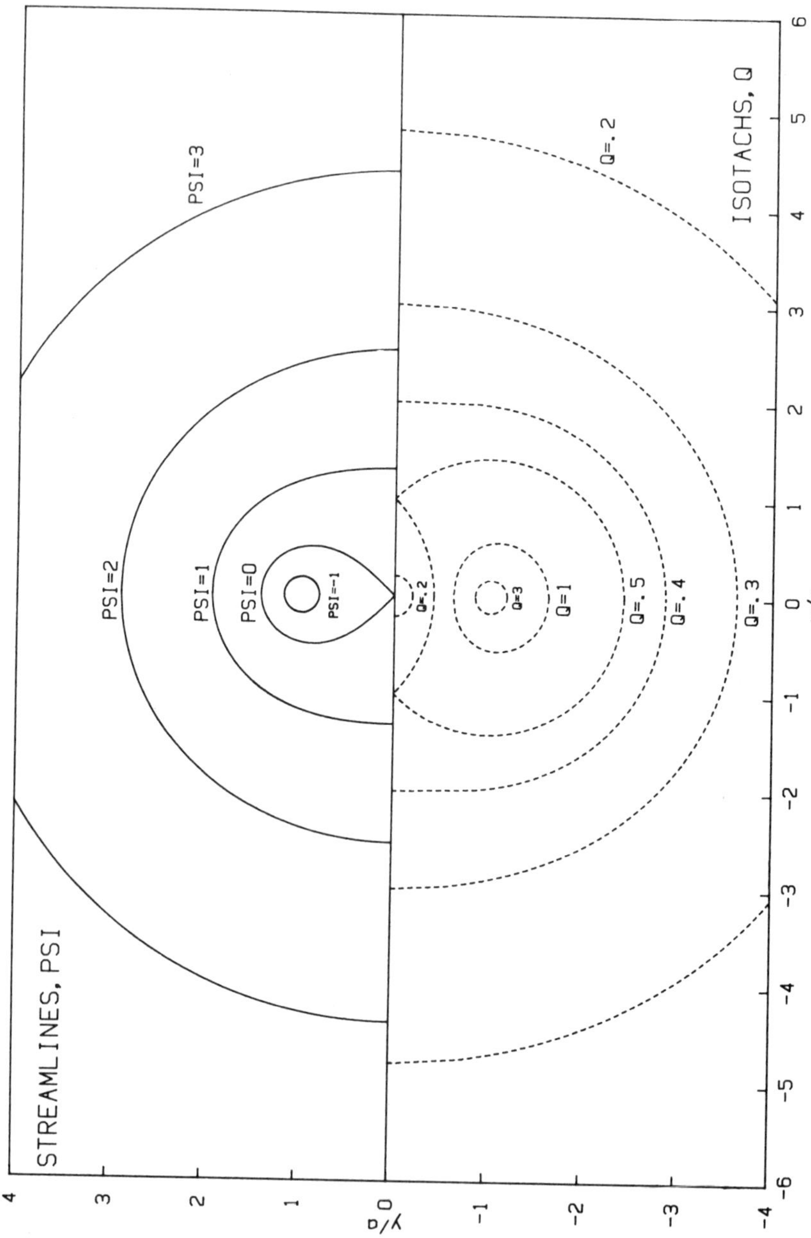

Figure 33 A vortex pair of identical rotation.

1.27 A VORTEX AND A SOURCE

Milne-Thomson (1960), page 360, gives the complex potential for the superposition of a source and a vortex at the origin

$$w = (-m + i\kappa) \log z \tag{81}$$

The stream function is written from (81) in dimensionless form as

$$PSI = \left(\frac{-m}{\kappa}\,\theta + \log\frac{r}{a}\right) \tag{82}$$

where PSI = ψ/κ. The parameter m/κ was chosen to be unity and the streamlines were plotted directly from (82) on Figure 34.

Equation (81) was differentiated to formulate the complex velocity and the speed from which the isotach equation is written as

$$Q = \frac{(1 + (m/\kappa)^2)^{1/2}}{r/a} \tag{83}$$

where $Q = qa/\kappa$. Isotachs are circles about the origin as shown on Figure 34. This flow pattern suggests the vanes of a pump or, if the source becomes a sink, the flow into a tornado.

1.28 A VORTEX BETWEEN WALLS

Milne-Thomson (1966), page 76, gives the complex potential for a vortex of strength κ midway between two parallel walls of spacing a.

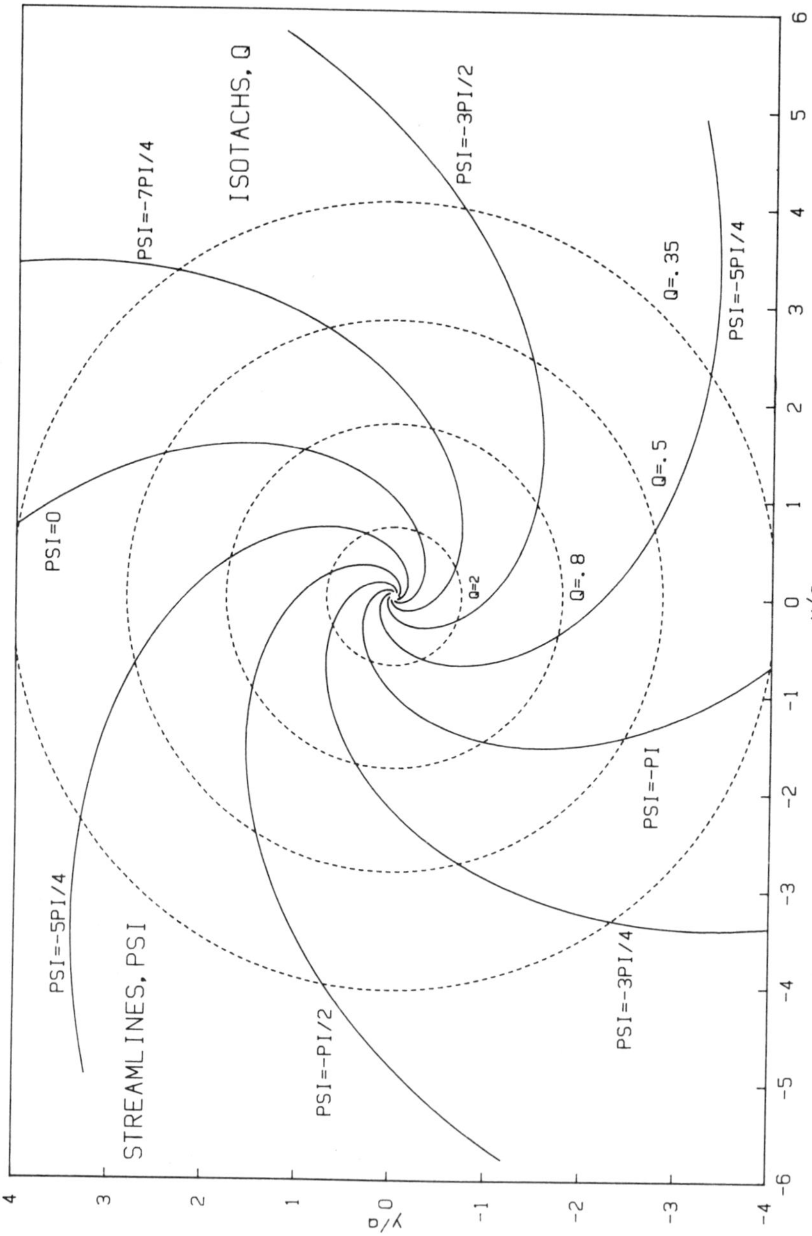

Figure 34 A source and a vortex.

$$w = i\,\kappa \log \tanh \frac{\pi z}{2a} \qquad (84)$$

There is a counterclockwise circulation about the origin. The stream function is written from (84) as

$$\psi = \frac{\kappa}{2} \log \frac{\cosh(\pi x/a) + \cos(\pi y/a)}{\cosh(\pi x/a) - \cos(\pi y/a)} \qquad (85)$$

The dimensionless stream function is defined as PSI = ψ/κ and streamlines were plotted from (85) on the upper half of Figure 35. Equation (85) was solved for y = f(x,PSI) using a Newton-Raphson technique.

Differentiating (84) to formulate the complex velocity and the speed yields the isotach equation

$$Q = \frac{\pi}{\sinh^2(\pi x/a) + \sin^2(\pi y/a)} \qquad (85)$$

where $Q = qa/\kappa$. Equation (85) was solved directly for x= f(y, Q) and the resulting isotachs are plotted on Figure 35.

1.29 A VORTEX IN A STREAM

When a uniform stream flowing from right to left is added to a counterclockwise vortex filament of strength κ at the origin (Milne-Thomson 1960, page 354), the resulting complex potential is

$$w = i\,\kappa \log z + Uz \qquad (87)$$

The dimensionless stream function is written directly from (87) as

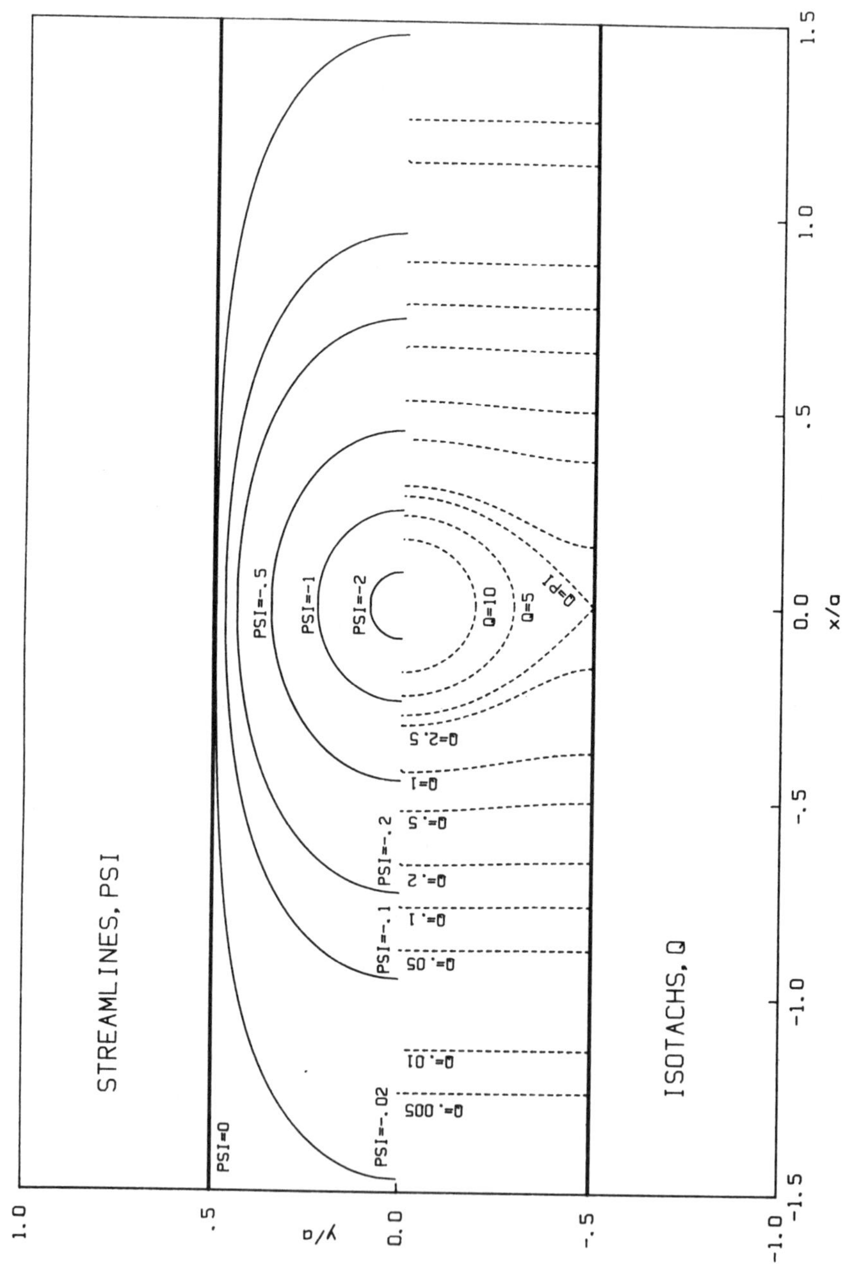

Figure 35 A vortex between two walls.

$$\text{PSI} = \frac{r}{a}\sin\theta + \frac{\kappa}{Ua}\log\frac{r}{a} \tag{88}$$

where PSI = ψ/Ua. Equation (88) is solved directly and streamlines are plotted on Figure 36 for the case of κ/Ua = 1. The streamline PSI = 0 can be interpreted as a hill whose height can be varied with the parameter κ/Ua.

Differentiating (87) to formulate the complex velocity and the speed yields the isotach equation

$$Q^2 = 1 + \left(\frac{\kappa}{Ua}\frac{a}{r}\right)^2 + 2\left(\frac{\kappa}{Ua}\frac{a}{r}\right)\sin\theta \tag{89}$$

where Q = q/U. Equation (89) was solved directly and the resulting isotachs are plotted on Figure 36.

1.30 THE VORTEX NOZZLE

The vortex shaped nozzle suggested by Robertson (1965) is discussed using the notation of Milne-Thomson (1960). A uniform flow from right to left is added to a clockwise vortex of strength κ at ia and a counter-clockwise vortex at -ia resulting in the complex potential

$$w = Uz + i\kappa\log\left(\frac{z + ia}{z - ia}\right) \tag{90}$$

The dimensionless stream function is written from (90) as

$$\text{PSI} = \frac{y}{a} + \frac{\kappa}{2Ua}\log\frac{(x^2 + y^2 + 2ay + a^2)}{(x^2 + y^2 - 2ay + a^2)} \tag{91}$$

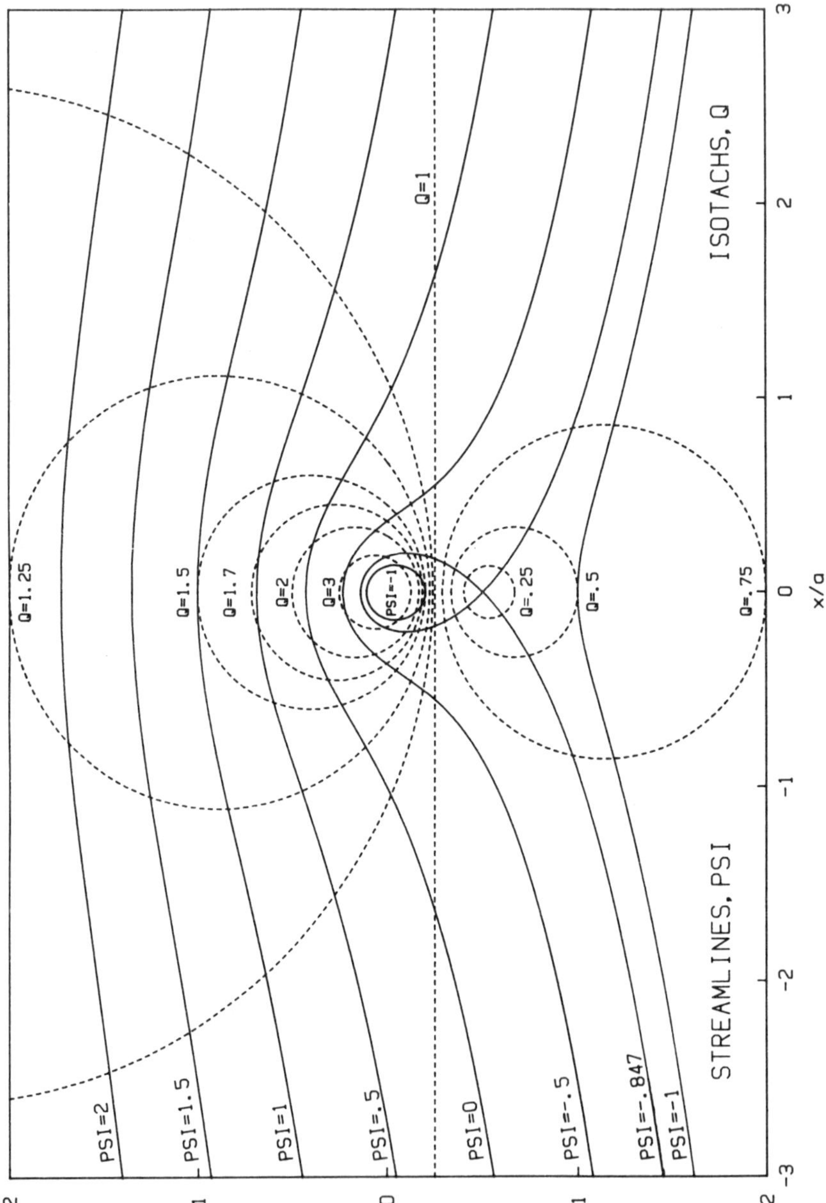

Figure 36 A vortex filament in a uniform stream.

where PSI = ψ/Ua. Equation (91) was solved directly for $x = f(y, PSI)$ with the parameter $\kappa/Ua = 2$. Streamlines are plotted on the upper half of Figure 37 with the arbitrary choice of the nozzle wall as PSI = $\pm$ 3.5. The nozzle contraction can be varied with the parameter κ/Ua.

Differentiating (90) to formulate the complex velocity and the speed yields the isotach equation

$$Q^2 = \frac{((z/a)^2 + 5)((\bar{z}/a)^2 + 5)}{((z/a)^2 + 1)((\bar{z}/a)^2 + 1)} \tag{93}$$

where $Q = q/U$ and $\kappa/Ua = 2$. The fourth order polynomial for the isotachs was solved numerically using a Newton-Raphson technique and the resulting isotachs are plotted on the lower half of Figure 37.

1.31 AN INFINITE ROW OF VORTICES

Milne-Thomson (1960), page 373, gives the complex potential for an infinite row of vortices each of strength κ separated by a along the x axis.

$$w = i\,\kappa \log \sin \frac{\pi z}{a} \tag{94}$$

The stream function is also given as

$$\psi = \frac{\kappa}{2} \log \frac{1}{2} (\cosh(2\pi y/a) - \cos(2\pi x/a)) \tag{95}$$

The dimensionless stream function is PSI = ψ/κ and (95) was solved directly for $x = f(y, PSI)$ with the resulting streamlines plotted on the upper half of Figure 38.

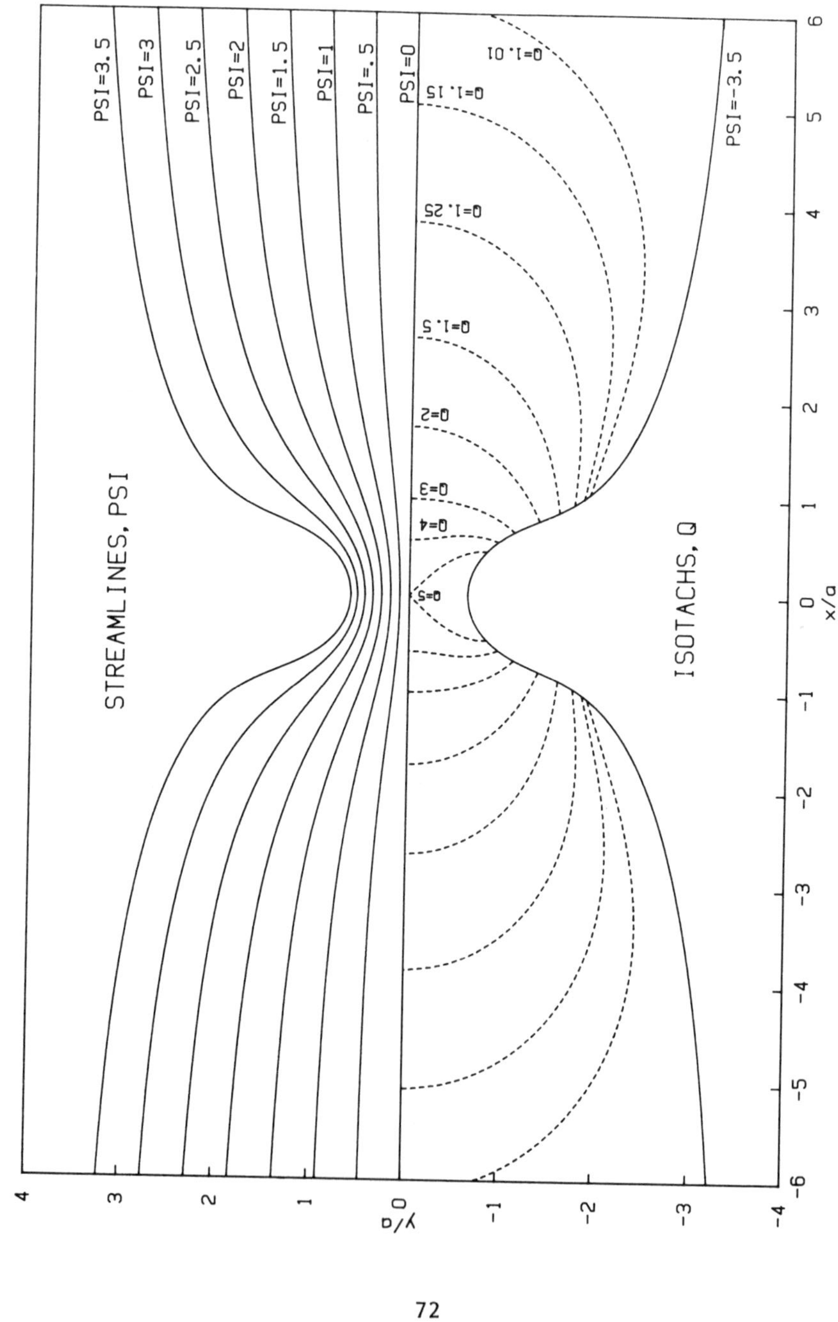

Figure 37 A vortex shaped nozzle.

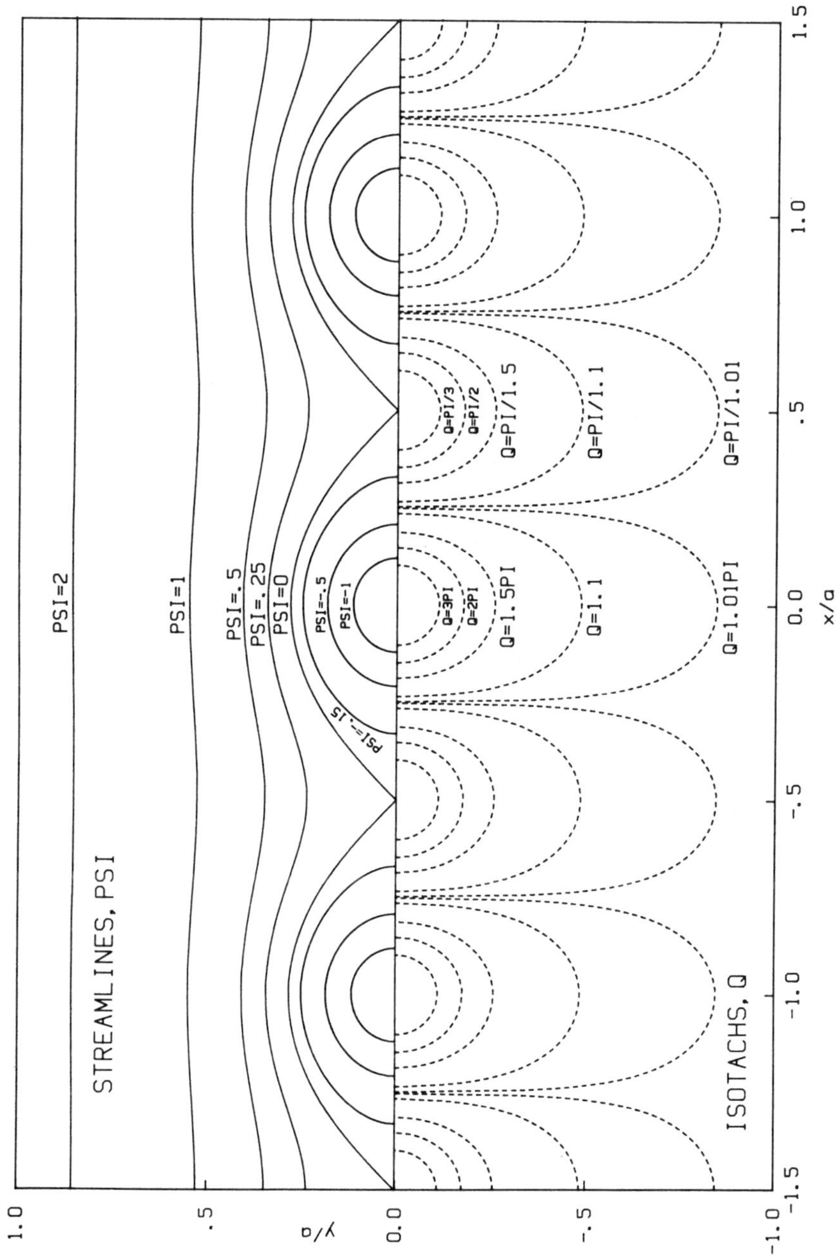

Figure 38 An infinite row of vortices.

Differentiating (94) to formulate the complex velocity and the speed yields the following isotach equation

$$Q^2 = \pi^2 \frac{(\cosh^2(\pi y/a) - \sin^2(\pi x/a))}{(\cosh^2(\pi y/a) - \cos^2(\pi x/a))} \tag{96}$$

where $Q = qa/\kappa$. Equation (96) was solved directly for $y = f(x, Q)$ and the resulting isotachs are plotted on the lower half of Figure 38. The speed approaches infinity at the vortices $0, \pm a, \pm 2a\ldots$ while a stagnation point exists at $\pm a/2$, $\pm 3a/2\ldots$ The whole vortex row is at rest and far from the vortex row, the speed approaches π.

1.32 KARMAN VORTEX STREET

Milne-Thomson (1960), page 375, gives the complex potential for two parallel infinite rows of opposite vortices κ and $-\kappa$. Vortices on the same row are separated by distance a, and the rows are separated by distance b.

$$w = i\kappa \log \sin \frac{\pi}{a}\left(z - i\frac{b}{2}\right) - i\kappa \log \sin \frac{\pi}{a}\left(z - \frac{a}{2} + i\frac{b}{2}\right) \tag{97}$$

The vortices in the upper row rotate counterclockwise while the vortices in the lower row have clockwise rotation. This configuration is known to be stable when $b/a = 0.281$. The steam function may be developed from (97) as

$$\mathrm{PSI} = \frac{1}{2} \log \left(\frac{\cosh 2\pi(y/a - k/2) - \cos 2\pi\, x/a}{\cosh 2\pi(y/a + k/2) - \cos 2\pi(x/a - 1/2)} \right) \tag{98}$$

where PSI = ψ/κ and k = b/a. Equation (98) was solved numerically using a Newton-Raphson technique for y = f(x, PSI) and streamlines for the upper row of vortices are plotted on the upper half of Figure 39. The PSI = 0 streamline separates the upper and lower vortex rows. There is an identical set of streamlines of opposite sign for the lower row of vortices which is obtained by shifting the present pattern to the right a distance x/a = 0.5 and rotating about the line y = 0. Far from the vortex row at large y/a, PSI $\rightarrow -k\pi$.

Differentiating (97) to formulate the complex velocity and the speed yields an equation for the isotachs.

$$\left(\frac{Q}{\pi}\right)^2 = \frac{\beta^2 - \alpha^2}{A^2C + B^2D} \tag{99}$$

where $C = \cosh^2(2\pi y/a) - \cos^2(2\pi x/a)$, $D = \cosh^2(2\pi y/a) - \sin^2(2\pi x/a)$, $Q = qa/\kappa$, $A = (\alpha^2 + \beta^2)/2$, $B = \alpha\beta$, $\alpha = \sinh(\pi k/2)$ and $\beta = \cosh(\pi k/2)$. Equation (99) was solved directly for x = f(y, Q) and the resulting isotachs are plotted on the lower half of Figure 39 for the lower row of vortices. This combined vortex street translates to the right with a speed $Q = \pi \tanh \pi k = 0.7078\pi$; thus, Figure 39 is an instantaneous view of the flow pattern. Far from the vortex row, the speed $Q \rightarrow 0$. The closely spaced streamline pattern is seen to produce a maximum speed of $Q = 2\pi$ at the points 0,0; ±.5,0; ±1,0;...

1.33 THE DISTRIBUTED SOURCE

Milne-Thomson (1960), page 227, gives the complex potential for a source of total output $2\pi m$ uniformly distributed along the real axis $0 \leqslant x \leqslant a$.

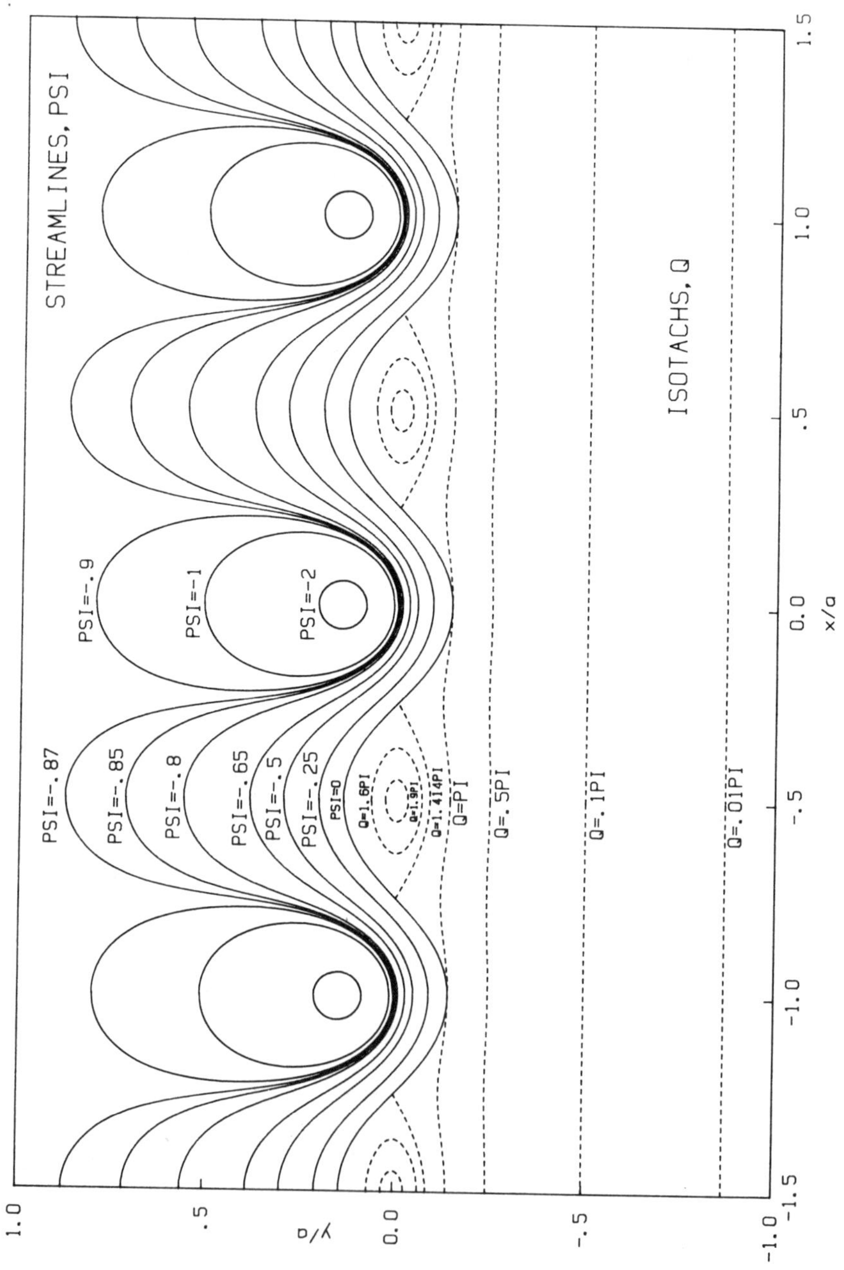

Figure 39 Karman vortex street.

$$w = m\left(\frac{z}{a}\log z - \frac{(z-a)}{a}\log(z-a)\right) \tag{100}$$

The dimensionless streamfunction is, corrected from Milne-Thomson (1960),

$$PSI = \frac{x}{a}(\theta_1 - \theta_2) + \theta_2 + \frac{y}{a}\log\left(\frac{r_1}{r_2}\right) \tag{101}$$

where PSI = ψ/m. The subscripts 1,2 refer to the bipolar coordinate description of the point x,y as shown on Figure 40. Equation (101) was solved numerically for the streamlines as plotted in the upper half of Figure 41.

Differentiating (100) to formulate the complex velocity and the speed yields the isotach equation

$$\left(\frac{r}{a}\right)^2(1 - \exp(-Q)) - \frac{2r}{a}\cos\theta + 1 = 0 \tag{102}$$

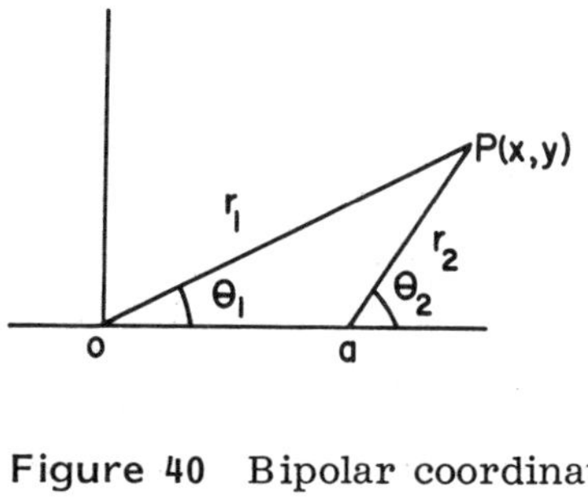

Figure 40 Bipolar coordinate system.

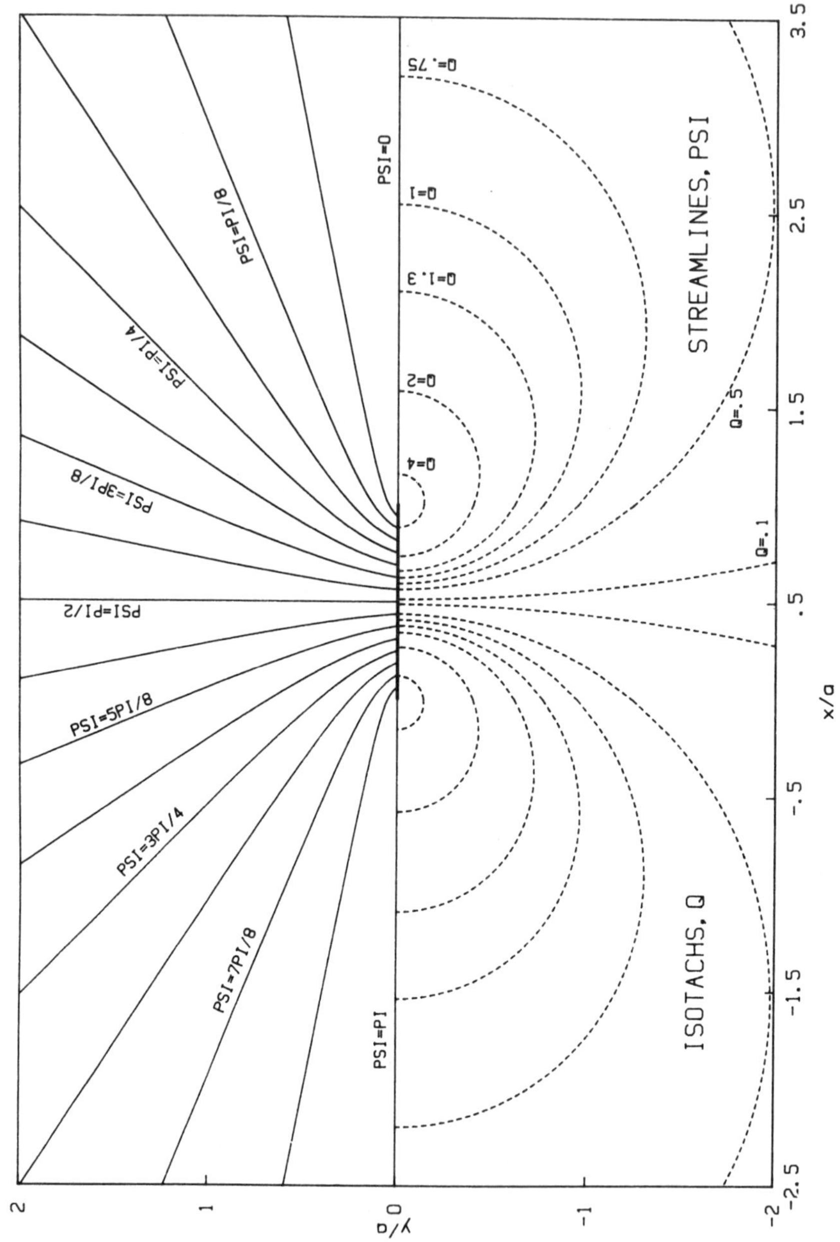

Figure 41 A distributed source.

where $Q = qa/m$. Equation (102) was solved for $r = f(Q, \theta)$ using the quadratic formula and isotachs are plotted on the lower half of Figure 41. The speed is zero at the center of the distribution and accelerates to infinity at the ends. This is in contrast to the three dimensional distributed source described in Section 2.7.

1.34 A DISTRIBUTED DOUBLET

In analogy to the distributed source described in Section 1.33, the complex potential for a doublet of total strength μ uniformly distributed along the real axis $0 \leqslant x \leqslant a$ may be written in the nomenclature of Milne-Thomson (1960) as

$$w = \frac{\mu}{a}\int_0^a \frac{d\xi}{(z - \xi)} \tag{103}$$

Upon integrating, equation (103) becomes

$$w = \frac{-\mu}{a} \log\left(\frac{z - a}{z}\right) \tag{104}$$

The dimensionless stream function $PSI = \psi a/\mu$ may be written from (104) as

$$PSI = (\theta_1 - \theta_2) \tag{105}$$

where the subscripts 1, 2 refer to the bipolar coordinates of Figure 40. Equation (105) was solved numerically using a Newton-Raphson technique for $y = f(x, PSI)$ and the streamlines are plotted on the upper half of Figure 42.

Equation (104) was differentiated to formulate the complex velocity and the speed yielding

$$Q = 1/(r_1 r_2) \tag{106}$$

where $Q = qa^2/\mu$. Equation (106) was solved numerically for the isotachs as plotted on the lower half of Figure 42. A minimum speed Q = 4 occurs on the distributed doublet with the speed going to infinity at the ends.

1.35 FLOW THROUGH A POROUS SCREEN

Taylor (1963) presents a solution for the flow through a porous screen which is modeled as a uniform flow U from left to right approaching a uniformly distributed source along the y axis at $x = 0$, $-b \leqslant y \leqslant b$. The components of velocity u,v at point P(x, y) are given in bipolar coordinates as

$$\frac{u}{U} = 1 - c(\theta_1 - \theta_2)$$

$$\frac{v}{U} = c \log\left(\frac{\sin\theta_1}{\sin\theta_2}\right) \tag{107}$$

where $c = \kappa/(4\pi + \pi\kappa)$ and κ is a dimensionless resistance coefficient related to the total source strength 2bm through the relation

$$\frac{m}{U} = \frac{\kappa b}{\pi(2 + \kappa/2)} \tag{108}$$

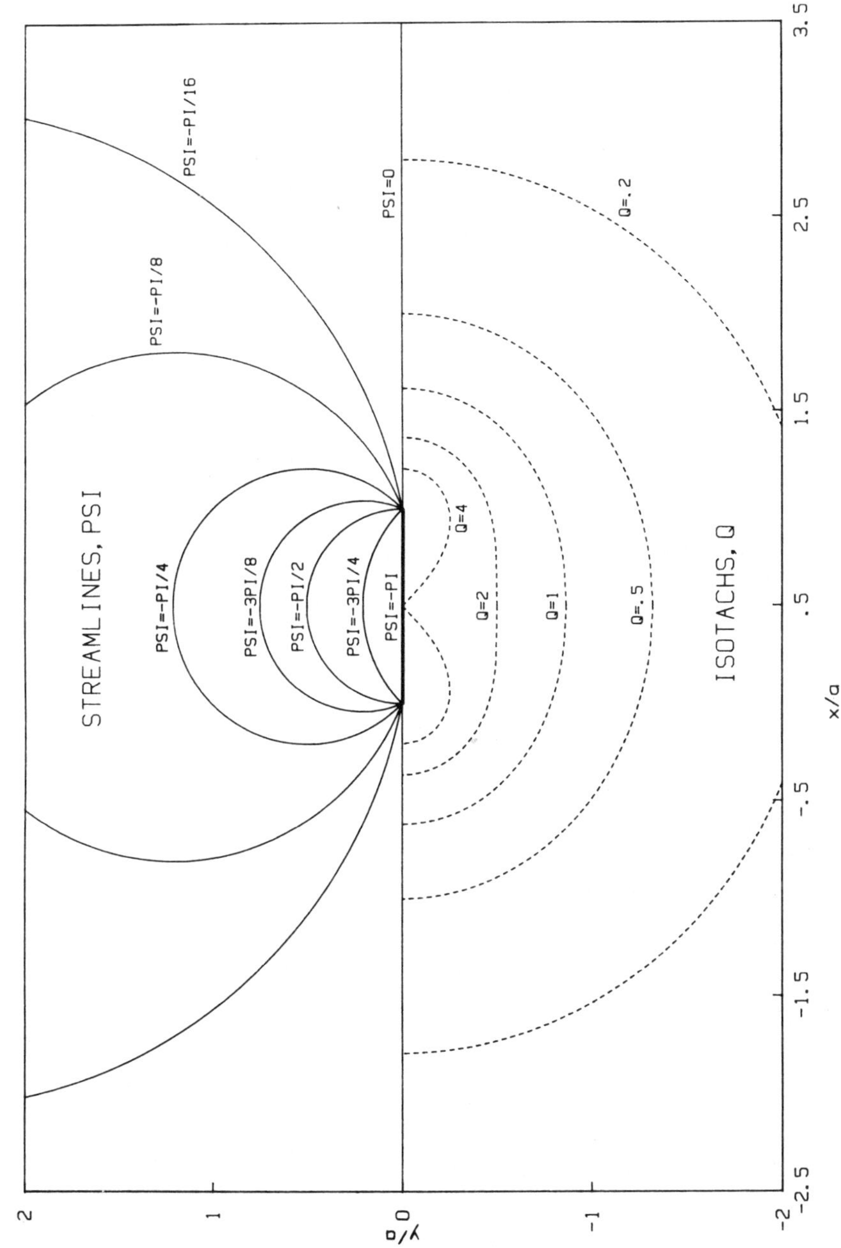

Figure 42 A distributed doublet.

The dimensionless stream function is PSI = ψ/Ub. The streamlines were constructed by numerically integrating along a streamline whose local shape is given by the velocity ratio v/u. Streamlines are shown on the upper half of Figure 43 for the case of κ = 4. The streamlines in the wake are not known because the model is only valid outside the wake. The amount of flow passing through the screen can be varied through the resistance coefficient κ.

The isotach equation is written from (107) as

$$Q^2 = \left(\frac{u}{U}\right)^2 + \left(\frac{v}{U}\right)^2 \qquad (109)$$

where Q = q/U. Equation (109) was solved numerically using a Newton-Raphson technique and the isotachs are plotted on the lower half of Figure 43. The dimensionless speed at the center of the screen is Q = 0.5 and the speed is seen to increase toward the edges of the screen.

1.36 FLOW PAST A VERTICAL PLATE

Milne-Thomson (1960), page 177, gives the complex potential for flow past a vertical plate of height 2c

$$w = U(z^2 + c^2)^{1/2} \qquad (110)$$

The streamline equation may be written from (110) as

$$\left(\frac{y}{c}\right)^2 = \frac{(x/c)^2 + 1 + PSI^2}{1 + (x/PSI)^2} \qquad (111)$$

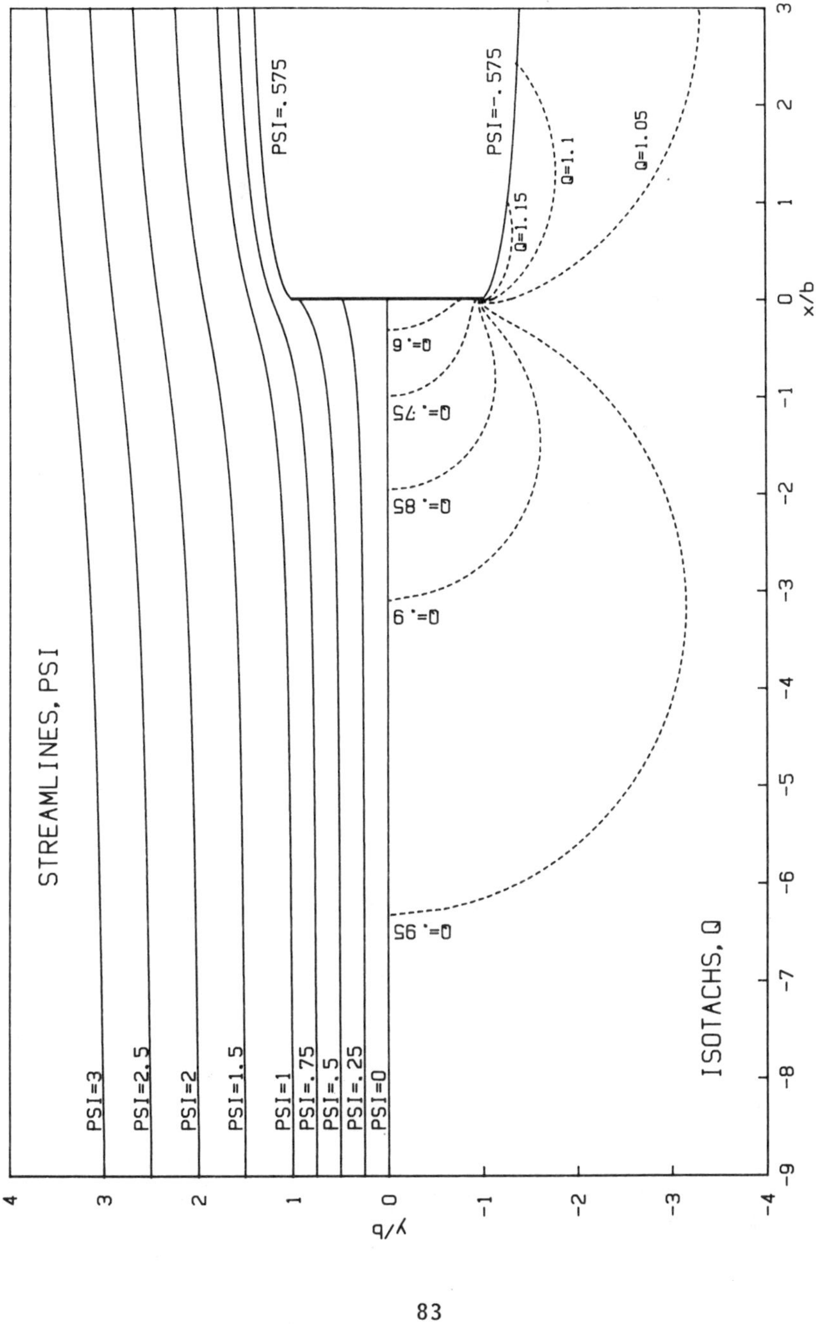

Figure 43 Flow through a porous screen.

where PSI = ψ/Uc. Dimensionless streamlines are plotted on the upper half of Figure 44. These nonseparated streamlines can be contrasted to those for separated flow past a plate in Section 1.48.

Differentiating equation (110) to formulate the complex velocity and the speed yields

$$Q^2 = z\,\bar{z}/(z^2\bar{z}^2 + \bar{z}^2 + z^2 + 1)^{1/2} \tag{112}$$

where $Q = q/U$. The polynomial in (112) was solved numerically using a Newton-Raphson technique and the isotachs are plotted on the bottom half of Figure 44. The flow stagnates at the center of the plate and accelerates along the plate to infinite speed at the sharp corner.

1.37 A TWO DIMENSIONAL PITOT TUBE

Streeter (1948), page 123, gives the complex potential for the uniform flow U around a two dimensional pitot tube of width 2b.

$$w = -\exp(w + z) \tag{113}$$

Equation (113) can be decomposed into

$$x = \frac{1}{2}\,\ell n\,((\phi)^2 + (\psi)^2) - \phi$$

$$y = \tan^{-1}(\psi/\phi) - \psi \tag{114}$$

Streamlines are generated from the set (114) in terms of $x = f(y, \psi)$ by numerically eliminating ϕ. The resulting stream-

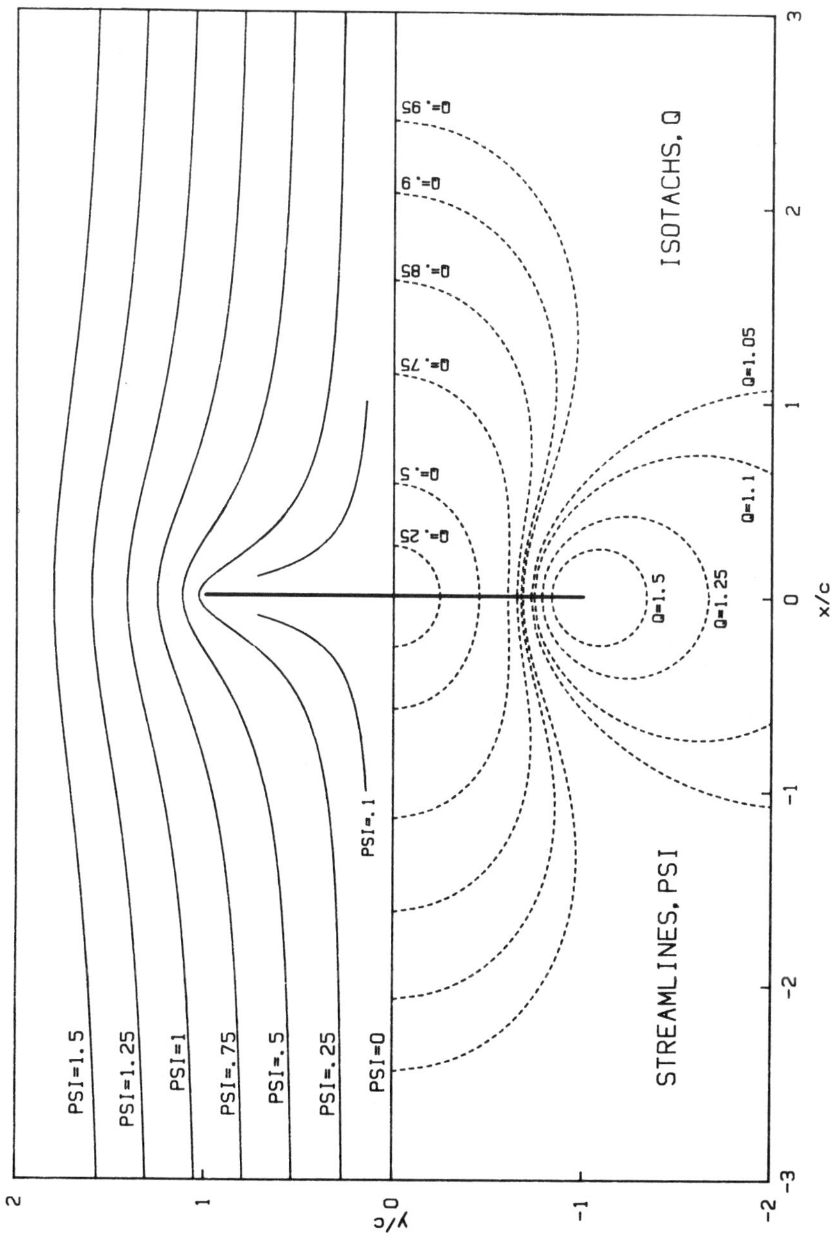

Figure 44 Nonseparated flow past a plate.

lines are plotted on the upper half of Figure 45 where PSI = ψ/Ub. The entrance plane to the pitot tube is at $x/b = 1/\pi$. Both the center streamline and the tube boundary have the value PSI = 0 indicating no net flow between them.

Differentiating (113) to formulate the complex velocity and the speed yields for the isotachs

$$Q^2 = \frac{\phi^2 + \psi^2}{\phi^2 + \psi^2 - 2\phi + 1} \tag{115}$$

where $Q = q/U$.

Equation (115) was solved for $\psi = f(Q, \phi)$ then transformed to the physical plane through the set of equations (114). Isotachs are plotted on the lower half of Figure 45. Stagnation of the flow occurs inside the tube and the speed tends to infinity at the sharp edges.

1.38 FLOW OVER A BOUNDED STEP

Milne-Thomson (1960), page 275, gives the complex potential for the flow over a bounded step of height (h-k) as

$$w = \frac{Uh}{\pi} \log \zeta \tag{116}$$

where ζ is linked to the physical plane through two transformations.

$$\zeta = \frac{b^2 - t^2}{1 - t^2} \tag{117}$$

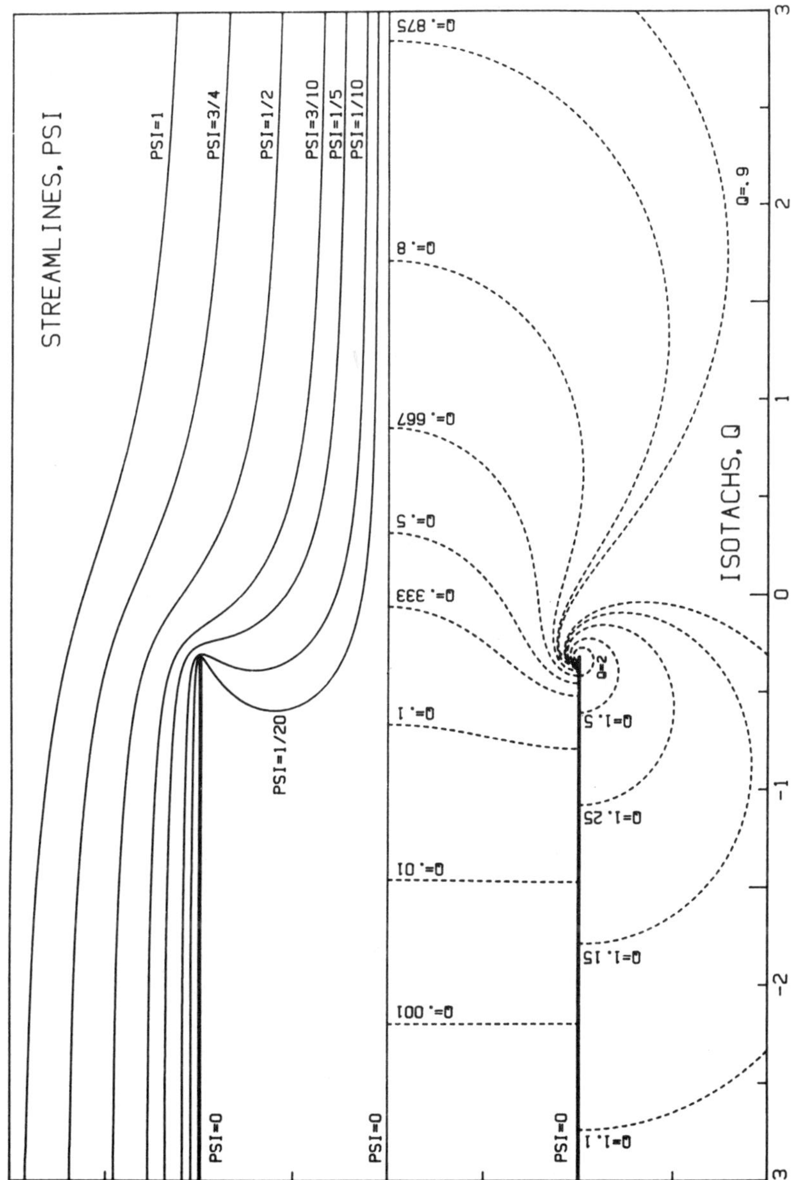

Figure 45 A two dimensional pitot tube.

and

$$z = \frac{h}{\pi}\left(\log \frac{1+t}{1-t} - \frac{1}{b}\log \frac{b+t}{b-t}\right) \qquad (118)$$

where b = h/k and both t and z are complex. Dimensionless streamlines are PSI = ψ/Uh and b = 2 for this case. If $\zeta = \rho$ exp (iγ), then the streamline equation may be written from (116) as

$$\text{PSI} = \gamma/\pi \qquad (119)$$

Thus, streamlines are rays in the ζ plane. Streamlines are plotted in the physical plane by transforming through the two mappings (117) and (118) as shown on Figure 46.

Differentiating (116) to formulate the complex velocity and the speed yields

$$Q^2 = t\,\bar{t} \qquad (120)$$

where Q = q/U. Thus, isotachs are circles in the t plane and they are plotted in the physical plane by transformation through (118) as shown on Figure 46. The flow stagnates at the bottom of the step accelerates to infinite speed at the sharp corner and then decelerates to Q = 2 far downstream of the step.

A listing of the software for this plot is in the Appendix.

1.39 FLOW OVER AN UNBOUNDED STEP

Milne-Thomson (1960), page 273, gives the complex potential for the flow over a step of height h.

$$w = \frac{Uh}{\pi}\,\zeta \qquad (121)$$

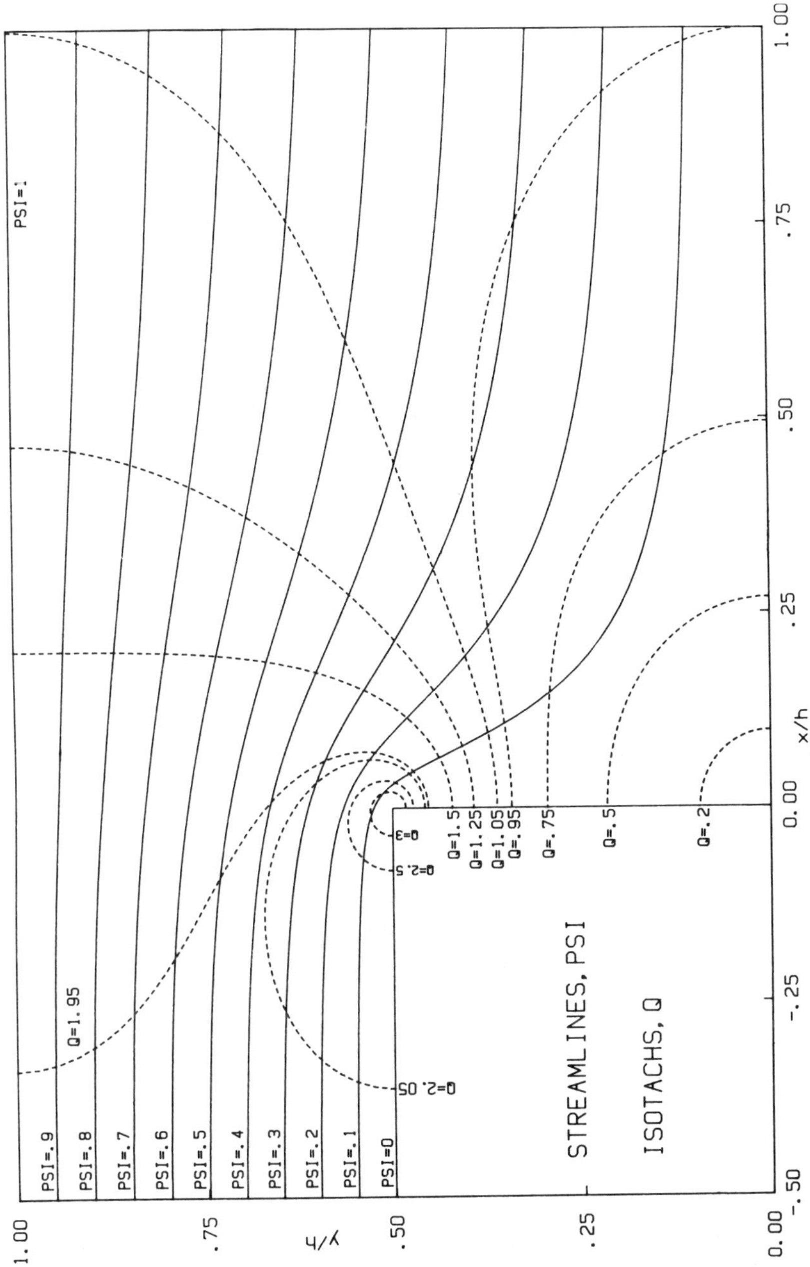

Figure 46 Flow over a bounded step.

where ζ is linked to the physical plane through the transformation

$$z = \frac{h}{\pi}\left((\zeta^2 - 1)^{1/2} + \cosh^{-1}\zeta\right) \tag{122}$$

When PSI = ψ/Uh and $\zeta = \rho \exp(i\gamma)$, the streamline equation may be written from (121) as

$$\text{PSI} = \frac{\rho}{\pi} \sin \gamma \tag{123}$$

Streamlines were generated from (123), transformed to the physical plane through (122) and are plotted on Figure 47.

Differentiating (121) to formulate the complex velocity and the speed yields

$$Q^4 = \frac{\rho^2 + 1 - 2\rho \cos \gamma}{\rho^2 + 1 + 2\rho \cos \gamma} \tag{125}$$

Equation (125) was solved for $\gamma = f(Q,\rho)$, transformed to the physical plane through (122) and plotted on Figure 47. The flow stagnates at the bottom of the step, accelerates to infinity at the sharp corner and returnes to Q = 1 far downstream of the step.

1.40 FLOW OVER A WAVY WALL

Michelson (1970), page 122, gives the potential function for the perturbation in the uniform flow U over a wavy wall of wave length L as

$$\phi(x,y) = Uh \exp(-\lambda y) \cos \lambda x \tag{126}$$

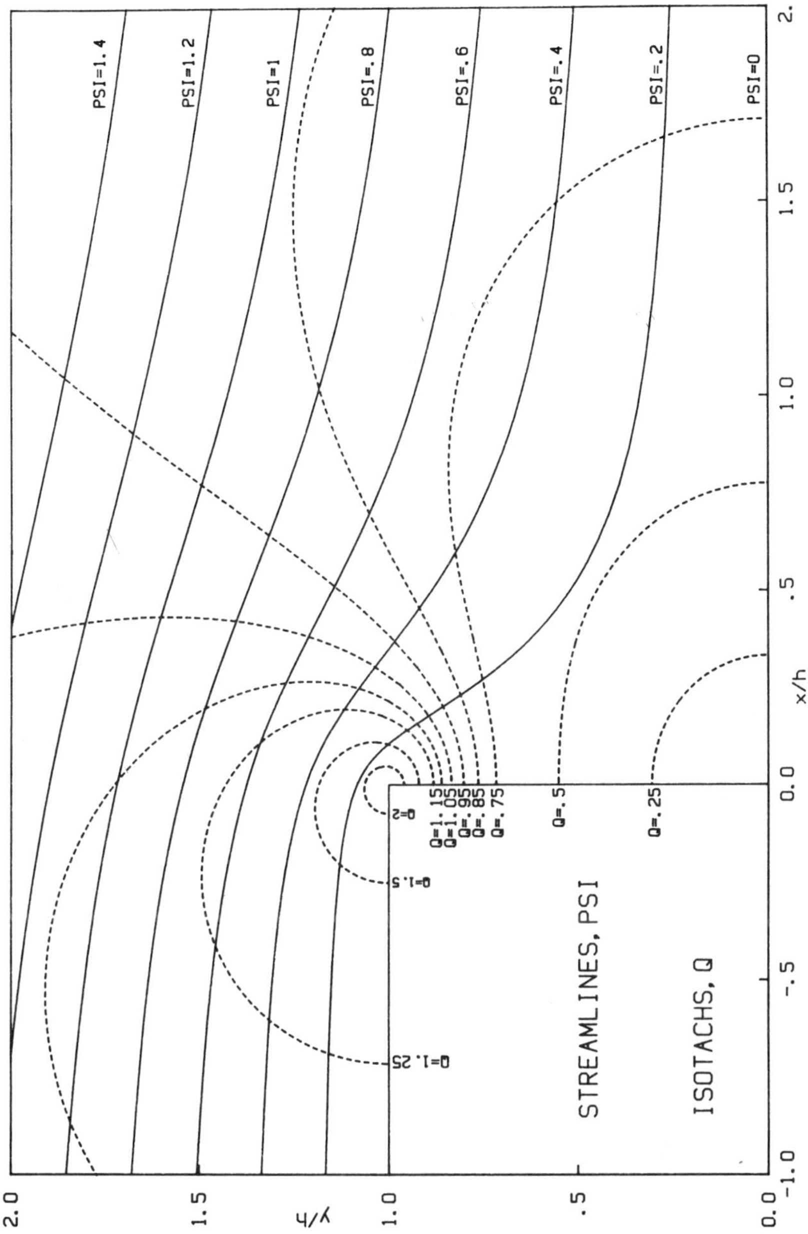

Figure 47 Flow over an unbounded step.

where the wavy wall is given by

$$y = h \sin \lambda x \tag{127}$$

with $\lambda = 2\pi/L$.

The velocity components for this flow are

$$u = U + \lambda U\, h \exp(-\lambda y) \sin \lambda x$$

$$v = \lambda U\, h \exp(-\lambda y) \cos \lambda x \tag{128}$$

The height to wave length ratio for the wall was chosen $h/L = 0.05$. Dimensionless streamlines PSI = ψ/UL were constructed by numerical integration along a streamline whose local slope is given by the velocity ratio v/u. Streamlines are shown on the left half of Figure 48.

The isotach equation is formulated from $q^2 = u^2 + v^2$ as

$$Q = 1 + 2\pi \frac{h}{L} \exp\left(-2\pi \frac{y}{L}\right) \sin 2\pi \frac{x}{L} \tag{129}$$

where $Q = q/U$. Isotachs plotted on the right half of Figure 48 are seen to be "in phase" with the wall perturbation.

1.41 THE FLAT PLATE AIRFOIL

Currie (1974), page 106, gives the complex potential for a flat plate airfoil of chord 4a at angle of attack α in a uniform flow U from left to right as

$$F(\zeta) = U\left(\zeta \exp(-i\alpha) + \frac{a^2}{\zeta} \exp(i\alpha)\right) + i2Ua \sin \alpha \log \frac{\zeta}{a} \tag{130}$$

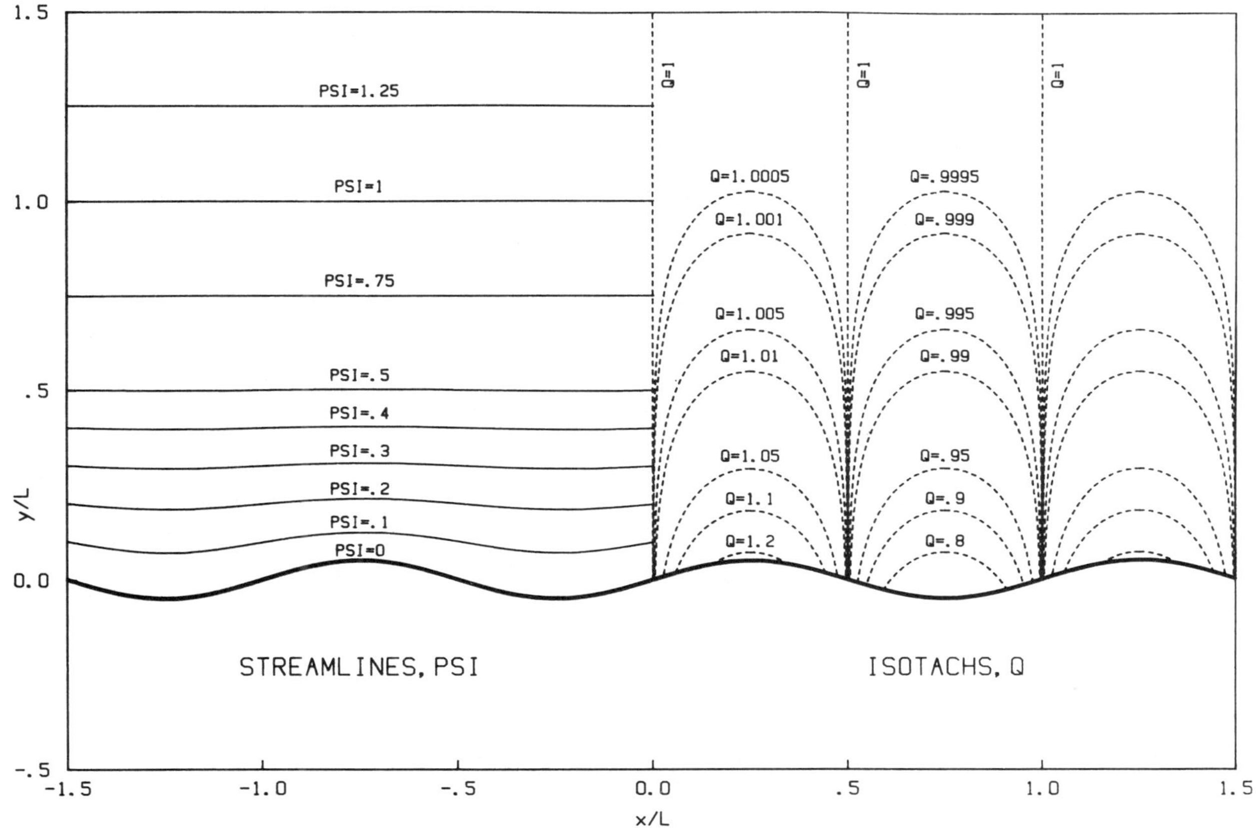

Figure 48 Flow over a wavy wall.

where ζ is linked to the physical plane through the Joukowski transformation

$$z = \zeta + \frac{a^2}{\zeta} \tag{131}$$

This flow has a clockwise circulation $\Gamma = 4\pi Ua \sin \alpha$ sufficient to satisfy the Kutta condition. The flat plate is generated by transforming a circle in the ζ plane into the z plane via equation (131). The streamline equation may be written from (130) as

$$\mathrm{PSI} = \left(\frac{\rho}{a} \sin(\nu - \alpha) + \frac{a}{\rho} \sin(\alpha - \nu)\right) + 2 \sin \alpha \log\left(\frac{\rho}{a}\right) \tag{132}$$

where $\zeta = \rho \exp(i\nu)$ and $\mathrm{PSI} = \psi/Ua$. Equation (132) was solved for $\rho = f(\nu, \alpha, \mathrm{PSI})$ using a Newton-Raphson technique and the results were transformed to the physical plane through (131). Streamlines for $\alpha = 15°$ are plotted on Figure 49. The stagnation point on the bottom of the plate is at $x/a = -2 \cos 2\alpha$.

The isotach equation is generated from (130) by formulating

$$Q^2 = \frac{1}{U^2} \frac{dF}{d\zeta} \frac{d\zeta}{dz} \frac{d\overline{F}}{d\overline{\zeta}} \frac{d\overline{\zeta}}{d\overline{z}} \tag{133}$$

resulting in

$$Q^2 = \frac{\rho^4 + 1 + (2\rho \sin \alpha)^2 + B(\rho^3 + \rho) - 2\rho^2 \cos 2(\nu - \alpha)}{A} \tag{134}$$

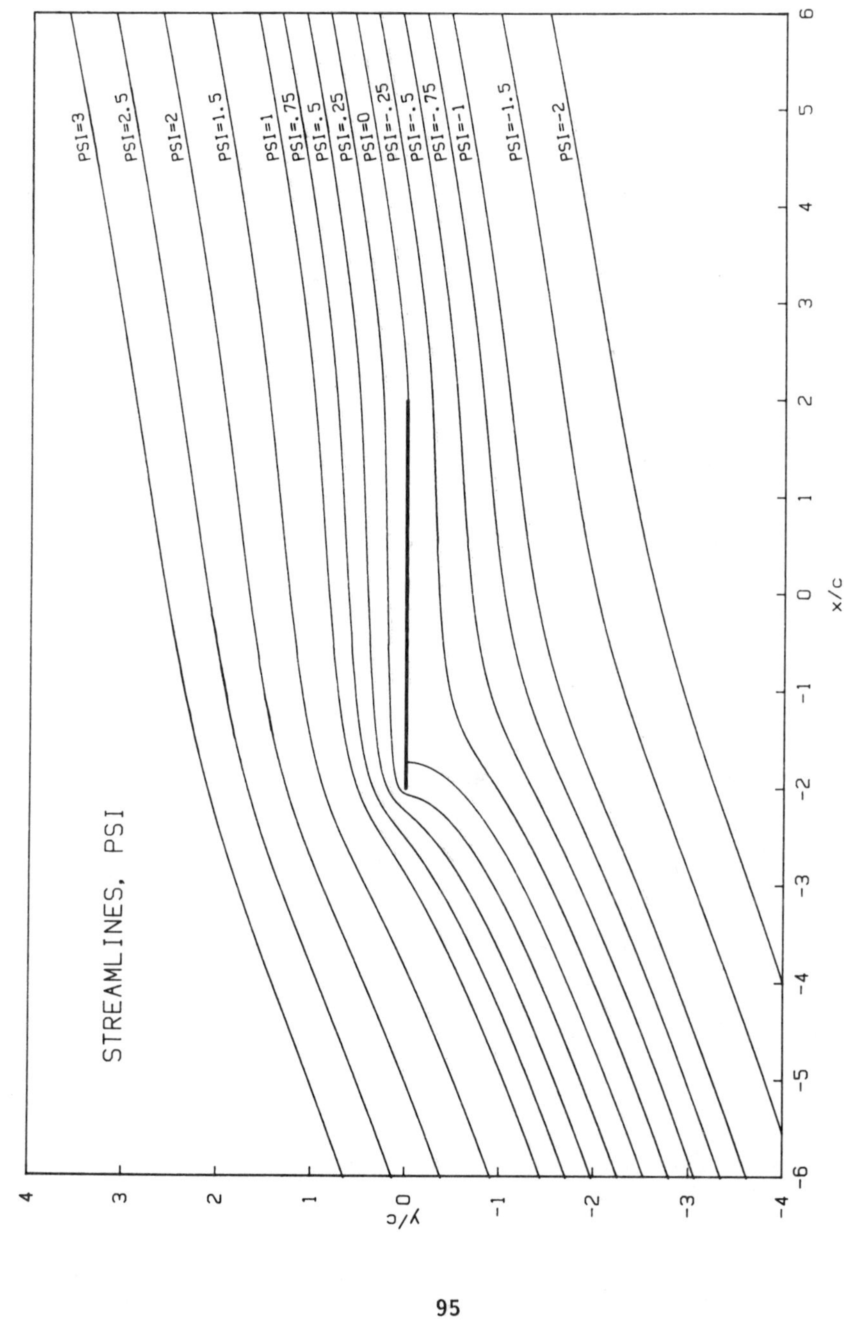

Figure 49 The flat plate airfoil, streamlines, alpha = 15°.

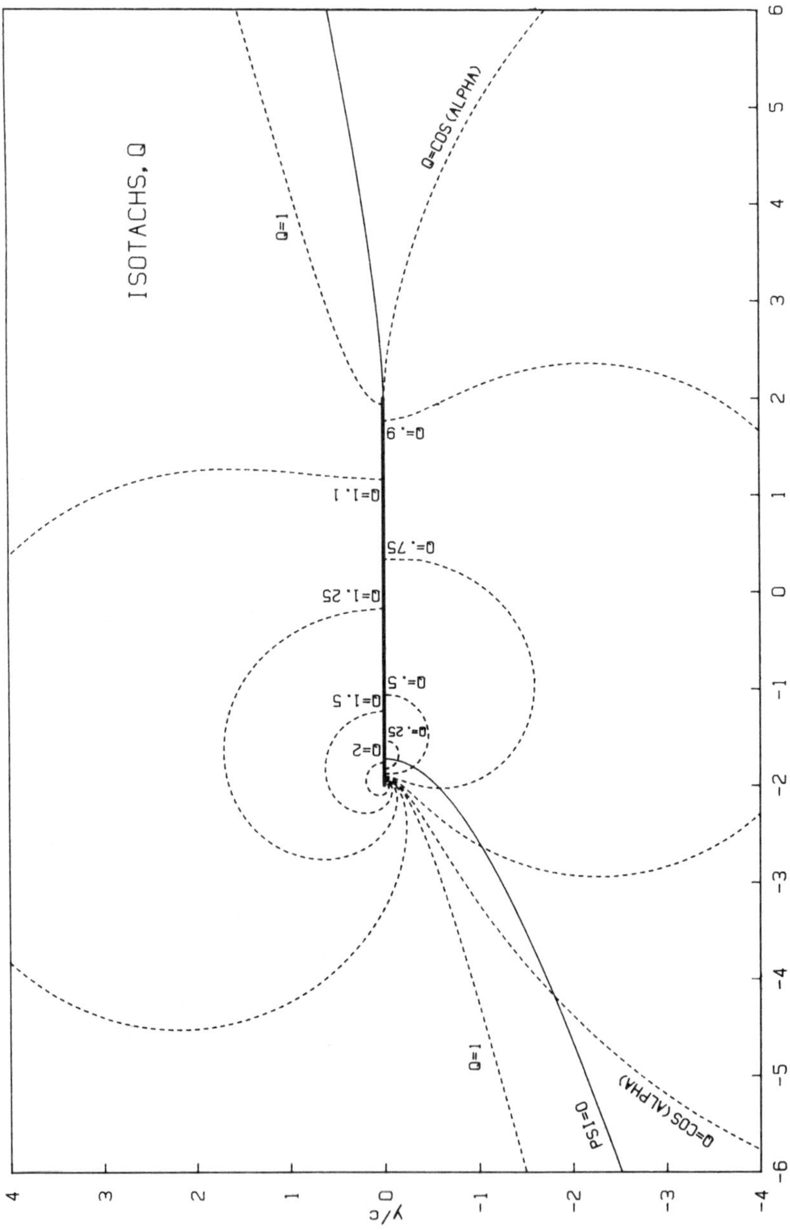

Figure 50 The flat plate airfoil, isotachs, alpha = 15°.

where $A = \rho^4 - 2\rho^2\cos 2\nu + 1$, $B = 4\sin\nu\sin(\nu - \alpha)$ and $Q = q/U$. Equation (134) was solved for $\rho = f(\nu, Q, \alpha)$ using a Newton-Raphson technique. Isotachs are plotted on Figure 50. The flow over the top and the bottom of the plate is separated by the stagnation streamline PSI = 0. The speed goes to infinity at the sharp leading edge and at the trailing edge $Q = \cos\alpha$.

The software for this flow is listed in the Appendix. In this example and all others using the Joukowski transformation, the flow over the cylinder was solved for lines of PSI and Q in the ζ plane and these coordinates were then transformed to the physical z plane.

1.42 THE SYMMETRIC JOUKOWSKI AIRFOIL; ALPHA = 0°

Currie (1974), page 109, gives the complex potential for a symmetric Joukowski airfoil of chord ℓ and thickness t which may be written for a flow of zero angle of attack as

$$F(\zeta) = U\left((\zeta + m) + \frac{a^2}{(\zeta + m)}\right) \qquad (135)$$

where $a = (\ell/4) + 0.77tc/\ell$ and $m = 0.77tc/\ell$. The airfoil is generated from a circle of radius a in the ζ plane transformed via equation (136) to the airfoil in the z plane.

$$z = \zeta + c^2/\zeta \qquad (136)$$

The dimensionless streamline equation may be written from (135) as

$$\mathrm{PSI} = \rho\sin\nu + \frac{\rho(1 + e)^2 \sin\nu}{\rho^2 + e^2 + 2\rho e\cos\nu} \qquad (137)$$

where PSI = ψ/Uc, $\zeta = \rho\exp(i\nu)$ and $e = 0.77t/\ell$. Equation (137) was solved for $\rho = f(\nu, PSI)$ using a Newton-Raphson technique and coordinates were transformed to the physical plane via equation (136). The resulting streamlines are plotted on the upper half of Figure 51.

The isotachs are generated from the relation $Q = dF/d\zeta \times d\zeta/dz \times 1/U$ which may be written from equations (135) and (136) as

$$Q = \left| \left(1 - \frac{(a/c)^2}{(\zeta/c + e)^2}\right)\left(\frac{(\zeta/c)^2}{(\zeta/c)^2 - 1}\right) \right| \tag{138}$$

where $Q = q/U$. Equation (138) was solved for $\rho = f(Q, \nu)$ using a Newton-Raphson technique on an eighth order polynomial in ρ. Isotachs are plotted on the lower half of Figure 51. The flow stagnates at the leading edge, accelerates to $Q = 1.2028$ at the maximum thickness and decelerates to $Q = 1/(1 + e)$ at the trailing edge.

1.43 THE SYMMETRIC JOUKOWSKI AIRFOIL; ALPHA = 15°

Currie (1974), page 109, gives the complex potential for a symmetric Joukowski airfoil of chord ℓ and thickness t in a uniform flow U at angle of attack α as

$$F(\zeta) = U\left((\zeta + m)\exp(-i\alpha) + \frac{a^2}{\zeta + m}\exp(i\alpha)\right) + \Lambda$$

$$\Lambda = \frac{i\Gamma}{2\pi}\log\left(\frac{\zeta + m}{a}\right) \tag{139}$$

where $a = (\ell/4) + 0.77tc/\ell$, $m = 0.77tc/\ell$ and $\Gamma = \pi U\ell(1 + 0.77t/\ell)\sin\alpha$. The circulation Γ is sufficient to satisfy the Kutta condition. The airfoil is generated from a circle of

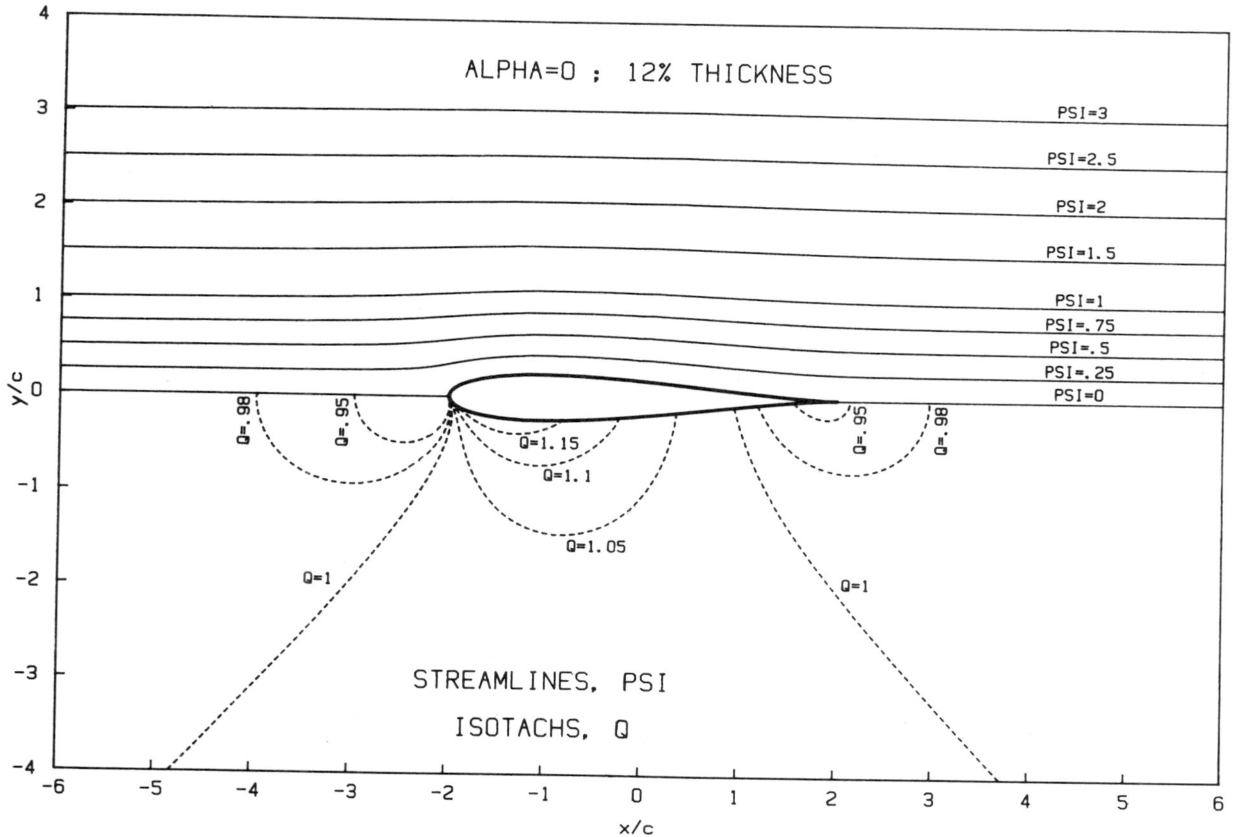

Figure 51 The symmetric Joukowski airfoil, alpha = 0°.

radius a in the ζ plane transformed via equation (136) to the airfoil in the z plane. The dimensionless streamline equation may be written from (139) as

$$\text{PSI} = \text{F2}(1 - 1/\text{F1}) + \text{A} \log \text{F1}$$

$$\text{F1} = (\rho^2 + 2e\rho\cos\nu + e^2)/(1 + e)^2$$

$$\text{F2} = \rho\sin(\nu - \alpha) - e \sin \alpha$$

$$\text{A} = 1 + e \sin \alpha \tag{140}$$

where PSI = ψ/Uc, e = 0.77t/ℓ and $\zeta = \rho\exp(i\nu)$. Equation (140) was solved for $\rho = f(\nu, \text{PSI}, \alpha)$, transformed to the physical plane, and plotted on Figure 52 for $\alpha = 15°$.

The isotach equation is generated from Q = dF/dζ × dζ/dz and written as

$$Q = \left| \left(\frac{(\zeta/c)^2}{(\zeta/c)^2 - 1} \right) \left(\exp(-i\alpha) - \frac{(1 + e)^2}{(\zeta/c + e)^2} \exp(i\alpha) + \frac{2i(1 + e)}{(\zeta/c + e)} \sin \alpha \right) \right| \tag{141}$$

Equation (141) was solved using a modified Newton-Raphson technique in which the slope of (141) was numerically calculated at a fixed ν, Q, α. The details are in the software listing in the Appendix. Isotachs are plotted on Figure 53. The flow stagnates on the bottom of the airfoil at $\nu = 208°$, accelerates to a maximum Q = 3.80 at $\nu = 177°$ and decelerates to Q = 0.8843 at the trailing edge.

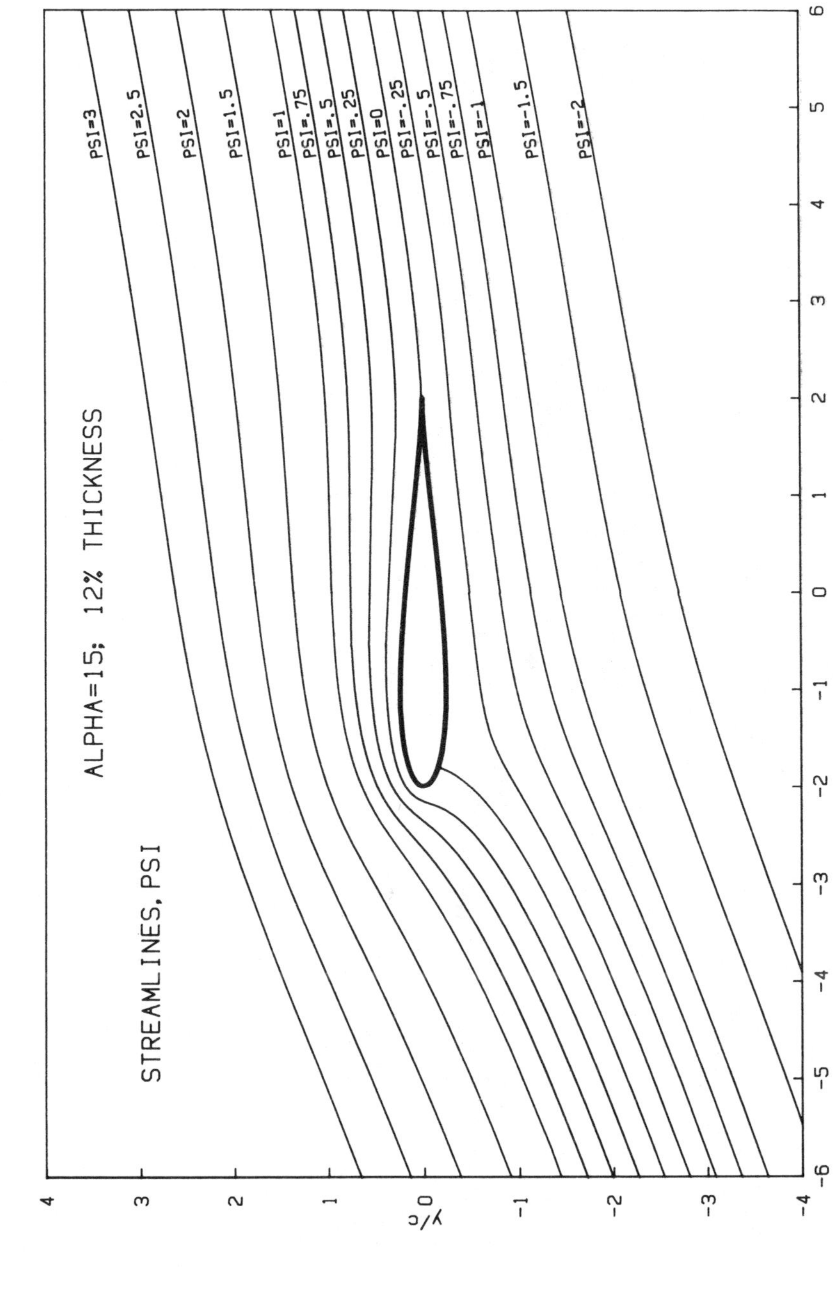

Figure 52 The symmetric Joukowski airfoil, streamlines, alpha = 15°.

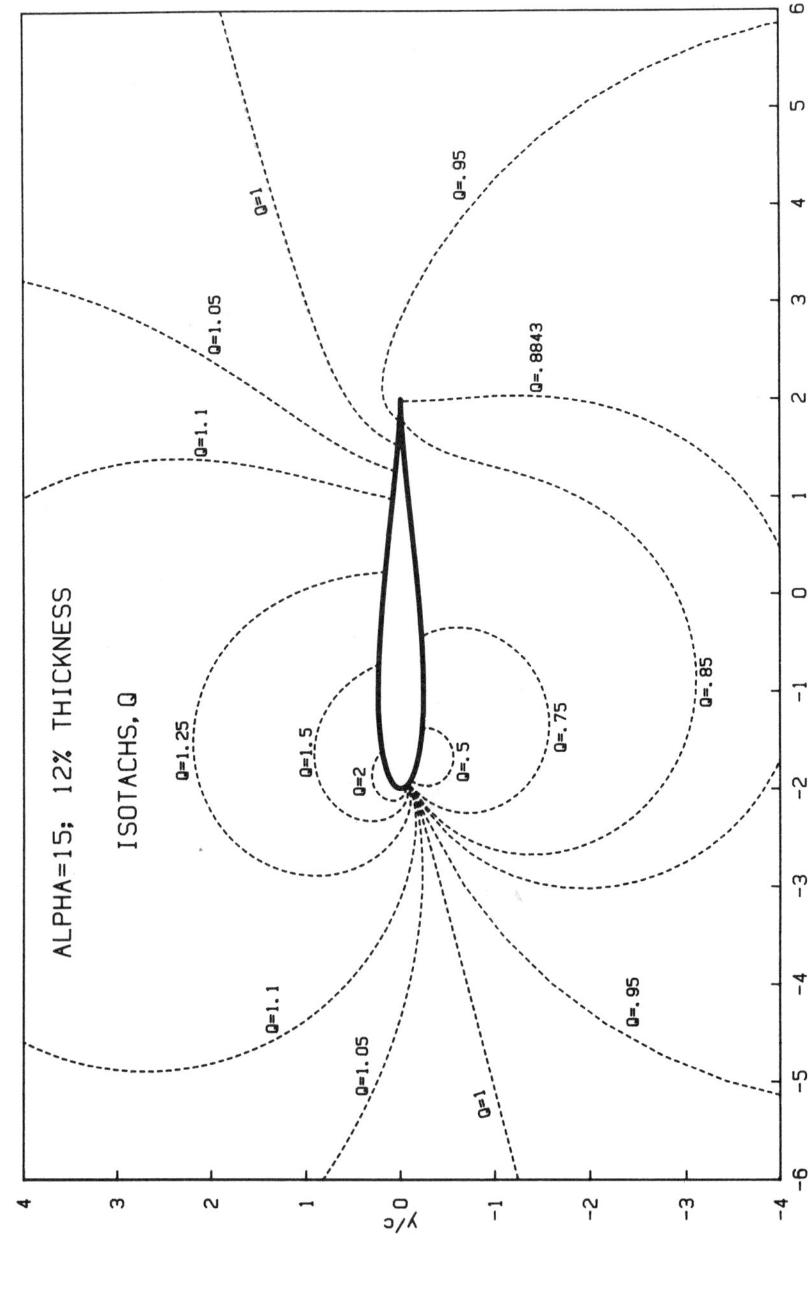

Figure 53 The symmetric Joukowski airfoil, isotachs, alpha = 15°.

1.44 THE CAMBERED JOUKOWSKI AIRFOIL

Currie (1974), page 118, gives the complex potential for a cambered Joukowski airfoil or chord ℓ, thickness t, and camber h in a flow U at angle of attack α as

$$F(\zeta) = U\left((\zeta - m \exp(i\delta)) \exp(-i\alpha) + \frac{a^2 \exp(i\alpha)}{\zeta - m \exp(i\delta)}\right) + \Lambda$$

$$\Lambda = \frac{i\Gamma}{2\pi} \log\left(\frac{\zeta - m \exp(i\delta)}{a}\right) \tag{142}$$

where $m \cos \delta = -0.77(tc/\ell)$, $m \sin \delta = h/2$, $a = (\ell/4) + 0.77\, tc/\ell$, $e = 0.77t/\ell$, and $\zeta = \rho\exp(i\nu)$. The circulation necessary to satisfy the Kutta condition is

$$\Gamma = \pi U \ell(1 + e) \sin(\alpha + 2h/\ell) \tag{143}$$

The airfoil is generated by transforming a circle of radius a centered at m exp ($i\delta$) in the ζ plane via equation (136) to the airfoil in the z plane. The dimensionless streamline equation may be written from (142) as

$$PSI = F1 - \frac{F1}{F2} + A \log(F2)$$

$$F1 = \rho\sin(\nu - \alpha) - e \sin(\delta - \alpha)$$

$$F2 = (\rho^2 - 2e\rho\cos(\nu - \delta) + e^2)/\gamma^2$$

$$A = \sin(\alpha + e \sin\delta) \tag{144}$$

where $PSI = \psi/Uc$ and $\gamma = (1 + e)$. Equation (144) was solved numerically using a Newton-Raphson technique for ρ =

f(ν, PSI, α) and then coordinates of the streamlines were transformed to the physical plane through the Joukowski transformation (136). Dimensionless streamlines for $\alpha = 0$, $t/\ell = 0.12$ and $h/\ell = .05$ are plotted on Figure 54. Streamlines for the same airfoil at $\alpha = 15°$ are plotted on Figure 55.

Milne-Thomson (1966), page 106, gives the complex potential for a cambered Joukowski airfoil of chord 4ℓ in a wind V at angle of attack α as

$$w = V\zeta\exp(i\alpha) + \frac{Va^2\exp(-i\alpha)}{\zeta - s} + 2\,aiV\sin\beta\log(\zeta - s) \tag{145}$$

where s is the center of the circle of radius a in the ζ plane. The ζ plane is linked to the z plane through the Joukowski transformation $z = \zeta + \ell^2/\zeta$. The angle β is determined from the equation $a\exp(i\beta) = (\ell + s)\exp(i\alpha)$. The dimensionless speed Q is also given as

$$Q = \left|\left(\exp(i\alpha) + \frac{a\exp(i\beta)}{\zeta - s}\right)\left(1 - \frac{a\exp(-i(\alpha+\beta))}{\zeta - s}\right)\right| \tag{146}$$

where $Q = q/V$. Equation (146) was solved with the same technique used for the solution of equation (141) in Section 1.43. Isotachs for $\alpha = 0$ are plotted on Figure 56 and for $\alpha = 15$ on Figure 57. These isotach plots are for the same airfoil shape of Figures 54 and 55. When $\alpha = 0$, the flow stagnates at the leading edge, accelerates to a maximum $Q = 1.342$ at $\nu = 123°$, and decelerates to $Q = .9077$ at the trailing edge. When $\alpha = 15°$, the flow stagnates at $\nu = 208.5°$, accelerates to a maximum $Q = 3.726$ at $\nu = 177°$, and decelerates to $Q = 0.8553$ at the trailing edge.

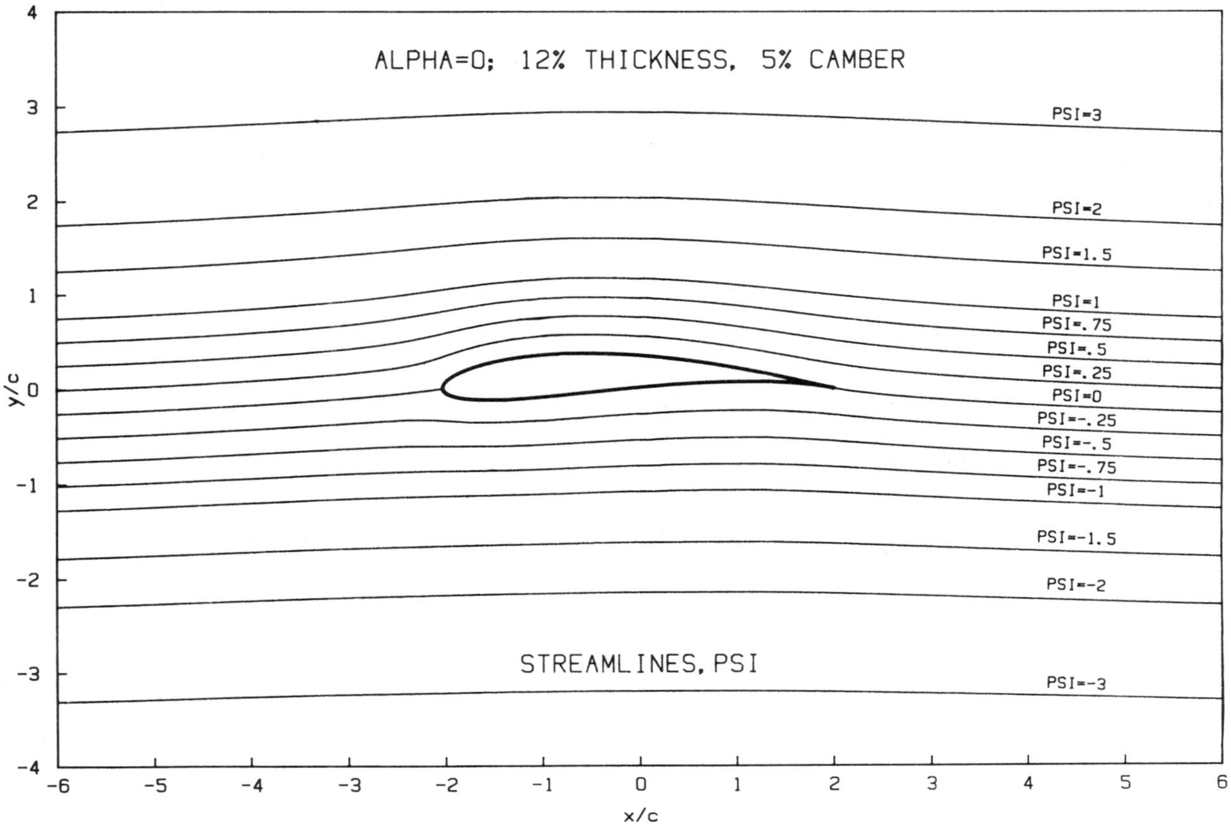

Figure 54 The cambered Joukowski airfoil, streamlines, alpha = 0°.

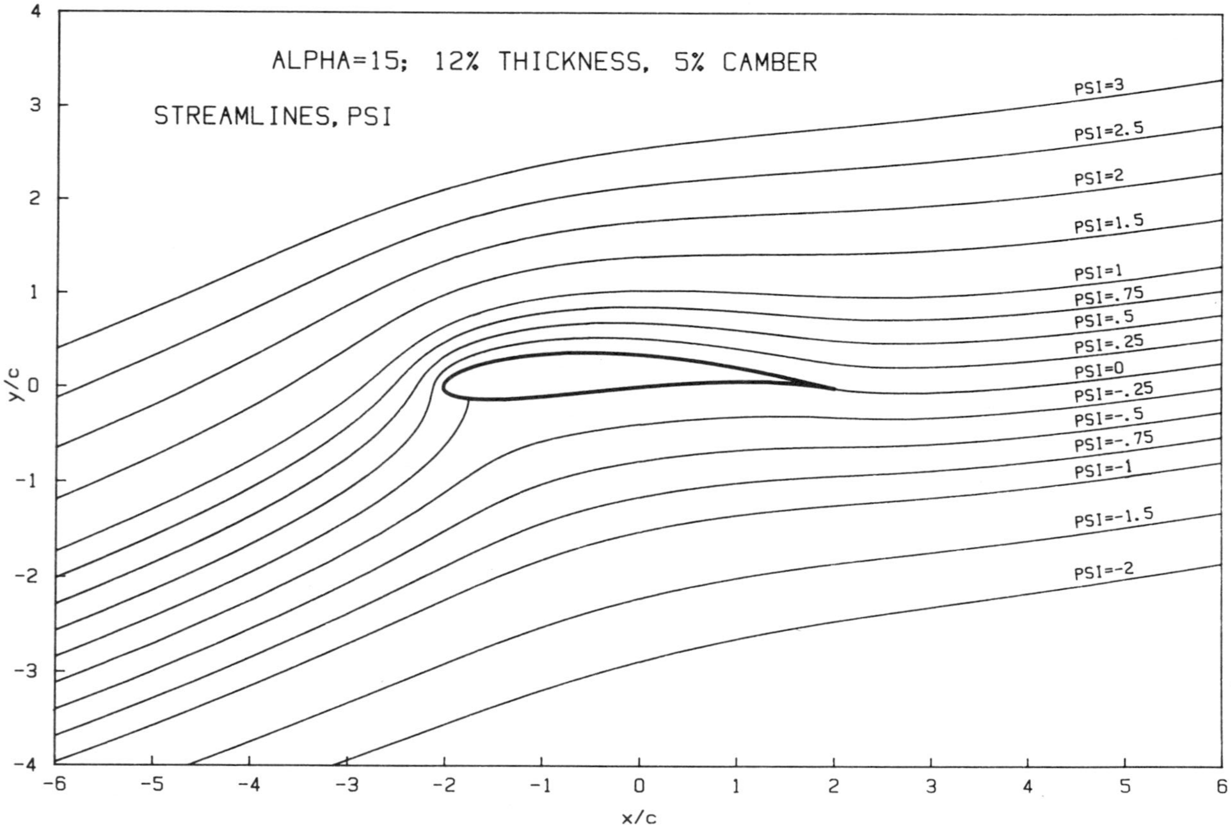

Figure 55 The cambered Joukowski airfoil, streamlines, alpha = 15°.

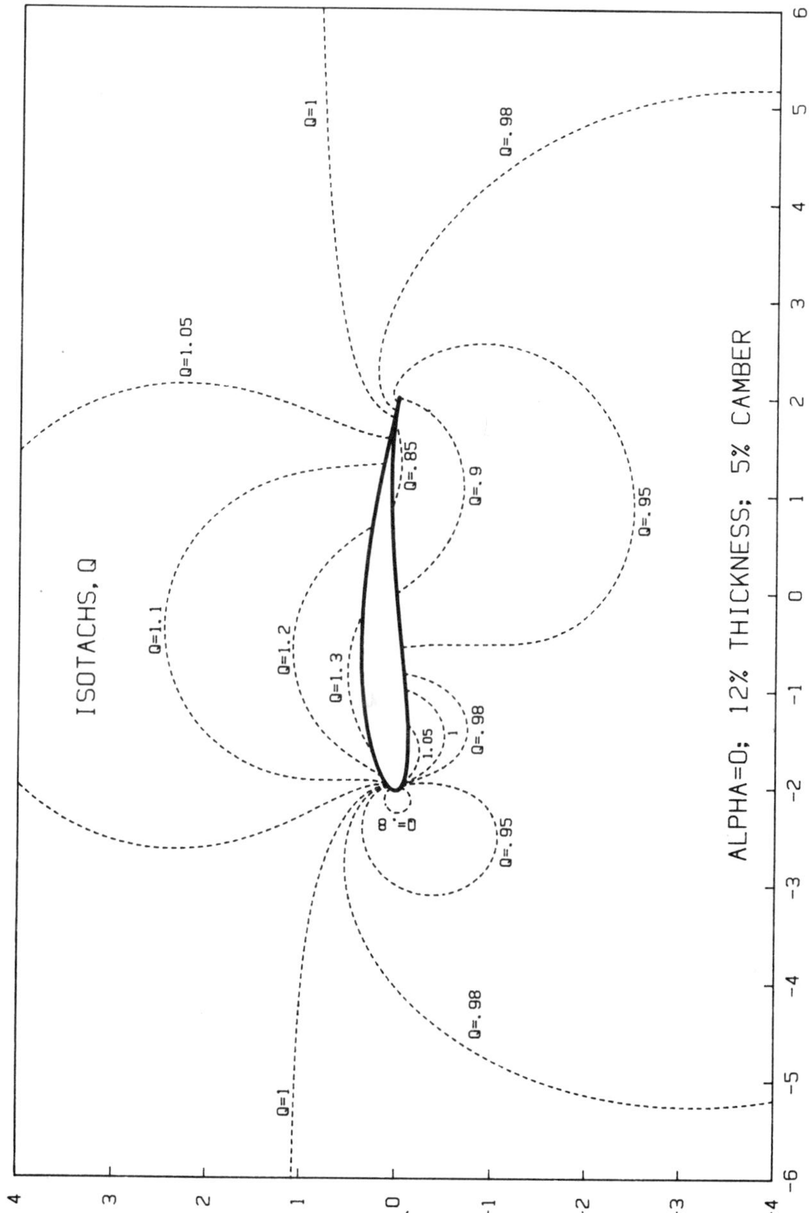

Figure 56 The cambered Joukowski airfoil, isotachs, alpha = 0°.

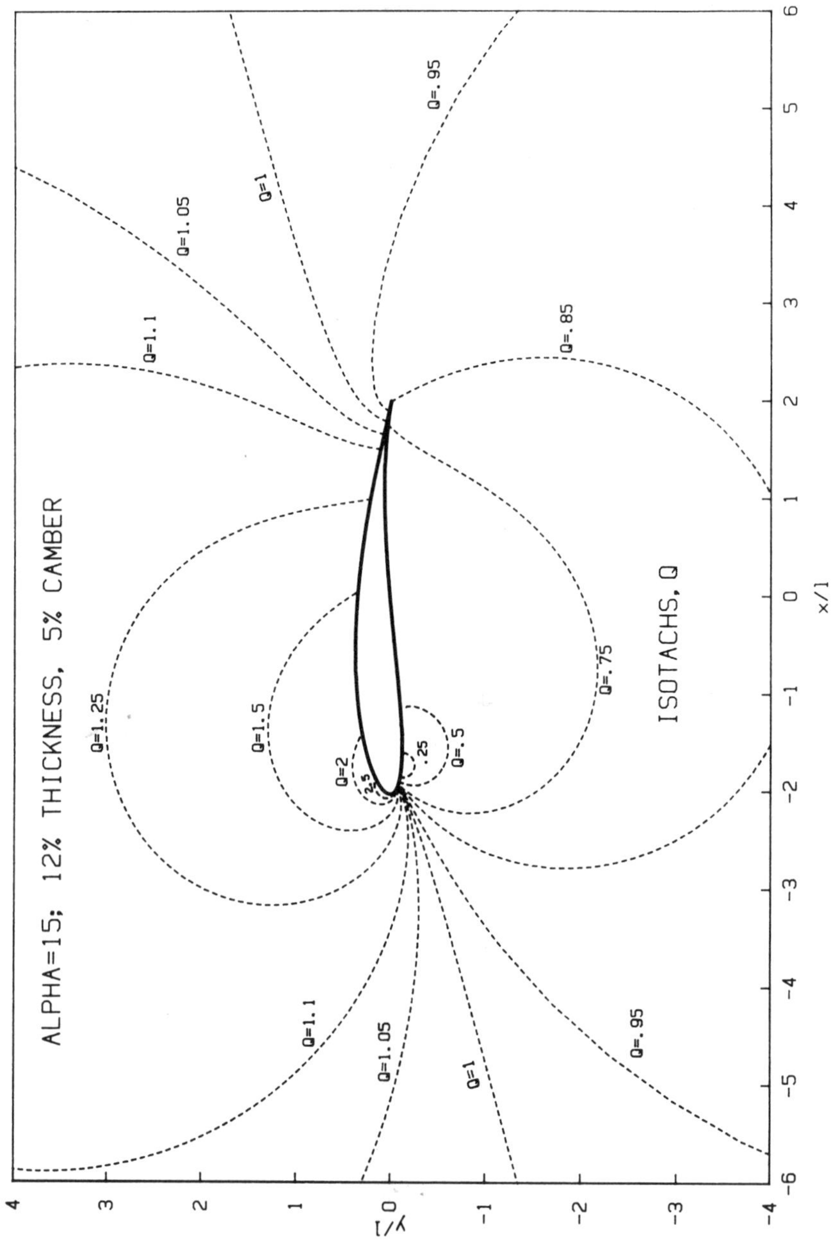

Figure 57 The cambered Joukowski airfoil, isotachs, alpha = 15°.

1.45 THE CIRCULAR-ARC AIRFOIL

Currie (1974), page 114, gives the complex potential for a circular arc airfoil of chord ℓ and camber height h in a flow U at angle of attach α as

$$\frac{F(\zeta)}{U} = \left((\zeta - im)\exp(-i\alpha) + \frac{a^2 \exp(i\alpha)}{(\zeta - im)}\right) + \frac{i\Gamma}{2\pi}\log\left(\frac{\zeta - im}{a}\right) \tag{147}$$

where $a = \ell/4$ and $m = h/2$. The circulation necessary to satisfy the Kutta condition is $\Gamma = \pi U\ell\sin(\alpha + 2h/\ell)$. The circular arc is generated by transforming a circle of radius a centered on the imaginary axis at $\zeta = im$ via the Joukowski transformation (136) to the physical plane using $c = \ell/4$. The dimensionless streamline equation may be written from (147) as

$$PSI = F2 - \frac{F2}{F1} + \sin(\alpha + e)\log F1$$

$$F1 = \frac{\rho^2 - 2e\rho\sin\nu + e^2}{1 + e^2} \tag{148}$$

$$F2 = \rho\sin(\nu - \alpha) - e\cos\alpha$$

where $\zeta = \rho\exp(i\nu)$, $e = 2h/\ell$ and $PSI = \psi/Uc$. Equation (148) was solved for $\rho = f(\nu, \alpha, PSI)$ using a Newton-Raphson technique and then coordinates were transformed to the physical plane using (136). Streamlines for $\alpha = 15°$ are shown on Figure 58 for a foil with 5% camber.

Differentiating equations (147) and (136) to generate $dF/d\zeta \times d\zeta/dz$ allows the isotach equation to be written as

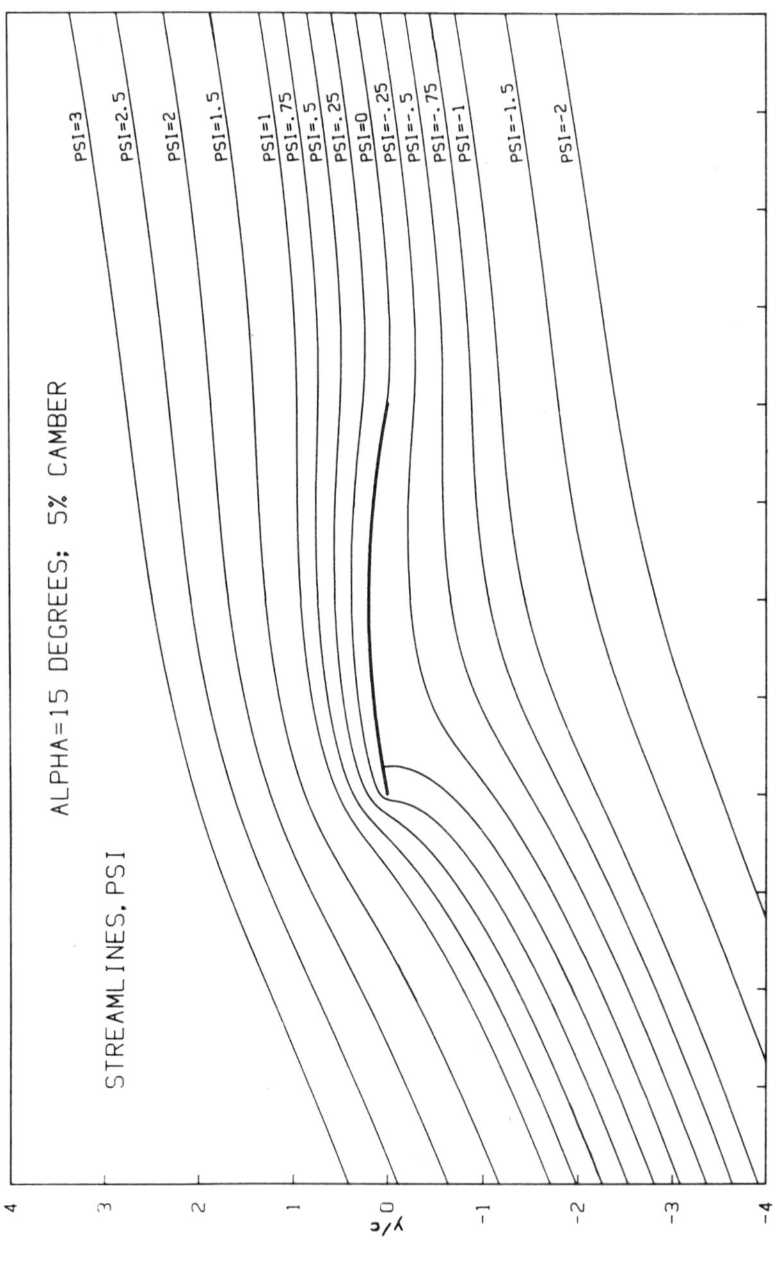

Figure 58 The circular-arc airfoil, streamlines, alpha = 15°.

$$Q = \left| \left(\frac{(\zeta/c)^2}{(\zeta/c)^2 - 1} \right) \left(\exp(-i\alpha) - \frac{(1 + e^2)\exp(i\alpha)}{(\zeta/c - ie)^2} + \frac{2i \sin(\alpha + e)}{(\zeta/c - ie)} \right) \right| \tag{149}$$

where $Q = q/U$. Equation (149) was solved using the same technique to solve (141) as described in Section 1.43. The resulting isotachs are plotted on Figure 59. The flow stagnates on the bottom of the foil, accelerates to infinity at the sharp leading edge and then decelerates to $Q = .9307$ at the trailing edge.

1.46 EFFLUX FROM AN ORIFICE

Kennard (1967), page 272, gives the complex potential for the efflux of a liquid from a two dimensional orifice of width 2a as

$$w = \frac{2a}{2 + \pi} \ln t \tag{150}$$

where the t plane is linked to the z plane through the transformation

$$z = \frac{2a}{2 + \pi} \left(t + (t^2 - 1)^{1/2} + \sin^{-1}(1/t) \right)$$

For simplicity, the constant speed s along the free streamline is taken to be $s = 1$. If the complex variable $t = \rho\exp(i\nu)$, then the streamline equation may be written from (150) as

$$PSI = \frac{2a}{2 + \pi} \nu \tag{151}$$

where $PSI = \psi/sa$. Thus, streamlines are radial lines in the t plane which are transformed to the z plane and plotted on the

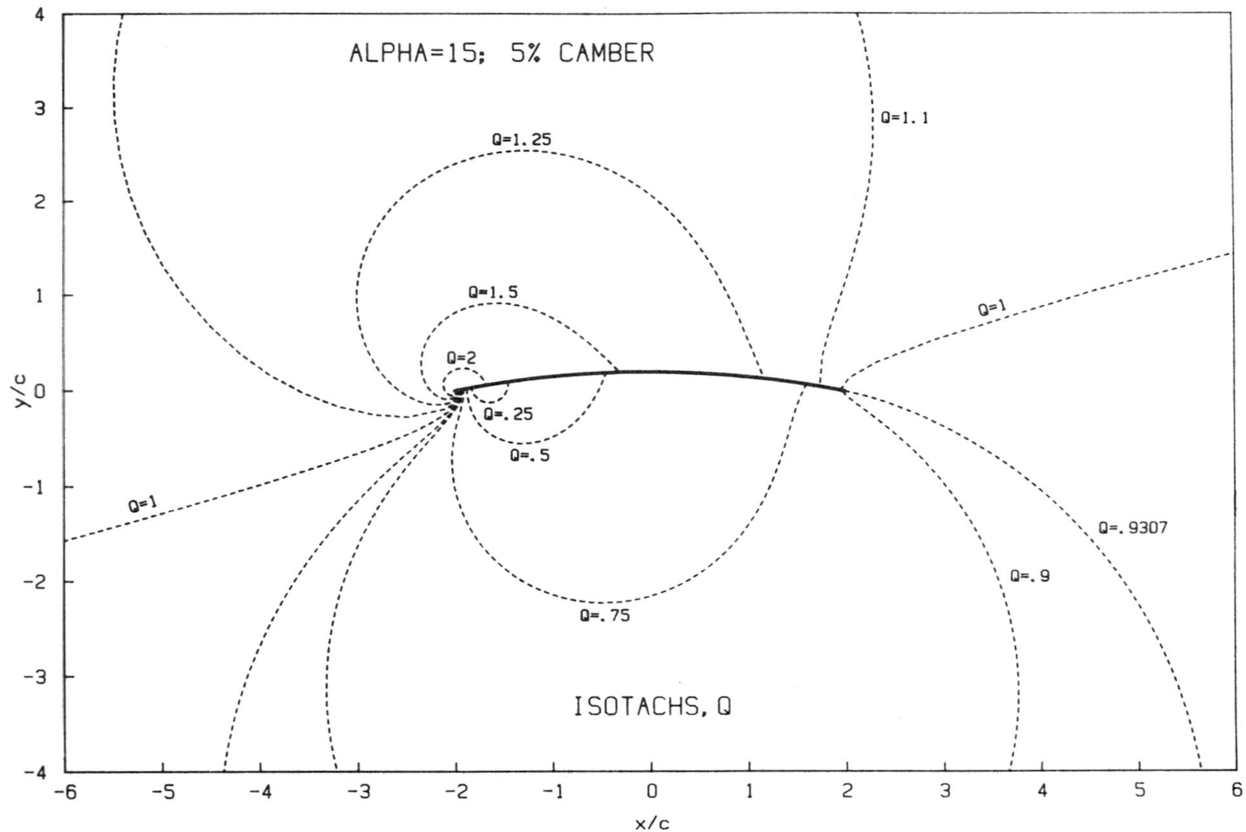

Figure 59 The circular-arc airfoil, isotachs, alpha = 15°.

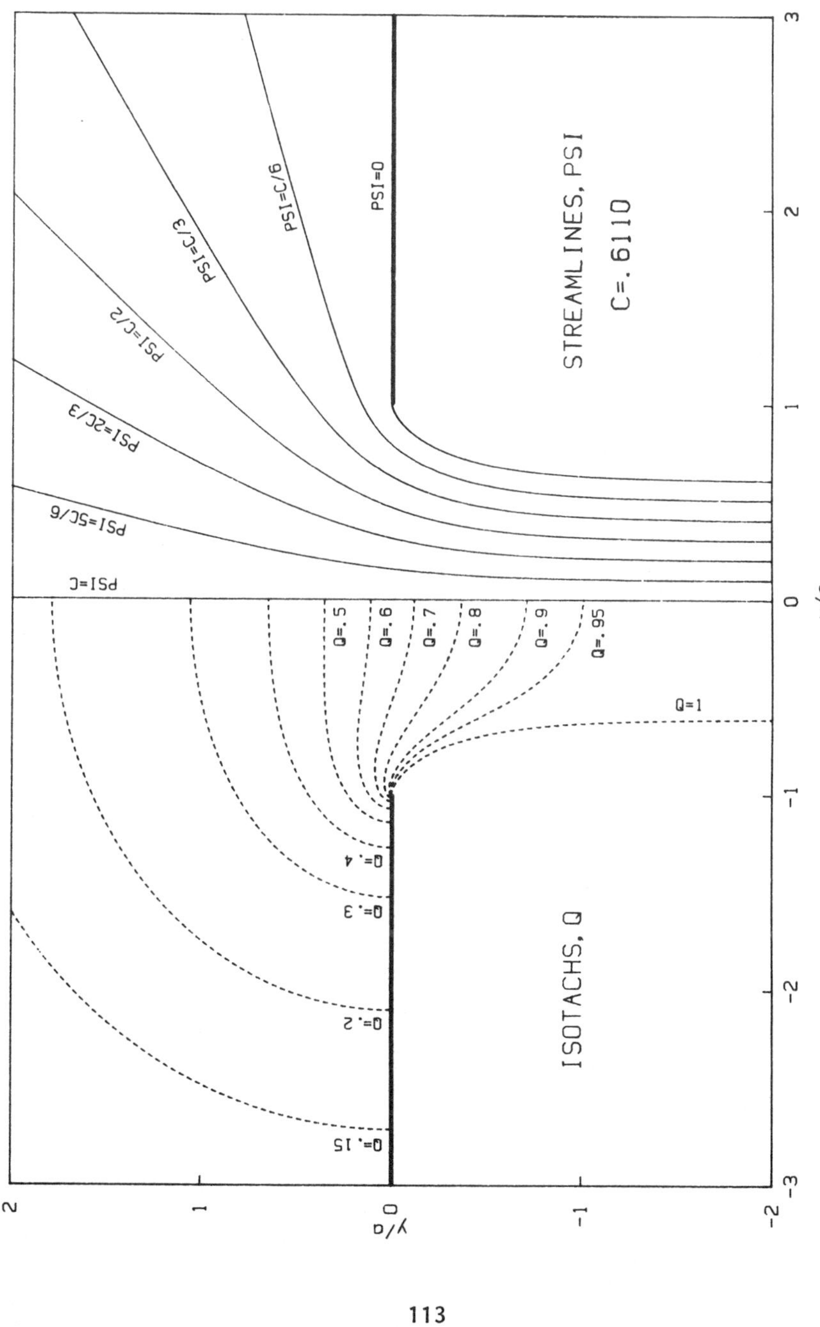

Figure 60 Efflux from an orifice.

right half of Figure 60. The streamline PSI = 0 is known as the free streamline because it is the boundary of the jet along which both the pressure and the speed are constant.

The complex velocity is also given as

$$-u + iv = (t + (t^2 - 1)^{1/2})^{-1} = 1/\zeta \tag{152}$$

thus $Q^2 = 1/\zeta\bar{\zeta}$. Isotachs are circles in the ζ plane which are transformed to the physical plane through (152) as plotted on the left half of Figure 60. The speed remains finite at the sharp edge of the orifice. The final width of the jet relative to the orifice width is the contraction ratio

$$C = \frac{\pi}{2 + \pi} = 0.611 \tag{153}$$

The dimensionless streamlines on Figure 60 are scaled with the contraction ratio.

1.47 BORDA'S MOUTHPIECE

Kennard (1967), page 278, gives the complex potential for flow into a canal of width 2a

$$w = \frac{a}{\pi} \ell n\ t \tag{154}$$

where the t plane is linked to the physical plane through the transformation

$$z = \frac{-a}{\pi} (t^2 + t(t^2 - 1)^{1/2} - \ell n\ t - \ell n(t + (t^2 - 1)^{1/2})) + \frac{a}{\pi} - ia \tag{155}$$

At the sharp edged entrance to the canal, the flow separates to form a free streamline along which both the speed and the pressure are constant. For simplicity, the speed s along the free streamline is taken to be s = 1. When the complex variable $t = \rho\exp(i\nu)$, the streamline equation may be written from (154) as

$$PSI = \nu/\pi \tag{156}$$

where $PSI = \psi/sa$. Thus, streamlines are radial lines in the t plane which are transformed to the z plane through (155) and plotted on the lower half of Figure 61.

The complex velocity is also given as

$$-u + iv = -\frac{1}{\zeta} = -(t + (t^2 - 1)^{1/2})^{-2} \tag{157}$$

Thus, $Q^2 = 1/\zeta\bar{\zeta}$. Isotachs are circles in the ζ plane which are transformed to the physical plane through (157) and (155) as plotted on the upper half of Figure 61. The speed entering the canal remains finite at the sharp edge in contrast with the flow of Figure 23 in which the speed goes to infinity at the sharp entrance. The final width of the jet in the canal is a; thus, the contraction ratio is 1/2.

The software developed to make this plot is listed in the Appendix.

1.48 SEPARATED FLOW PAST A PLATE

Kennard (1967), page 281, gives the complex potential for the flow past a plate of width ℓ as

$$w = -\frac{\ell}{\pi + 4}\,\frac{1}{t^2} \tag{158}$$

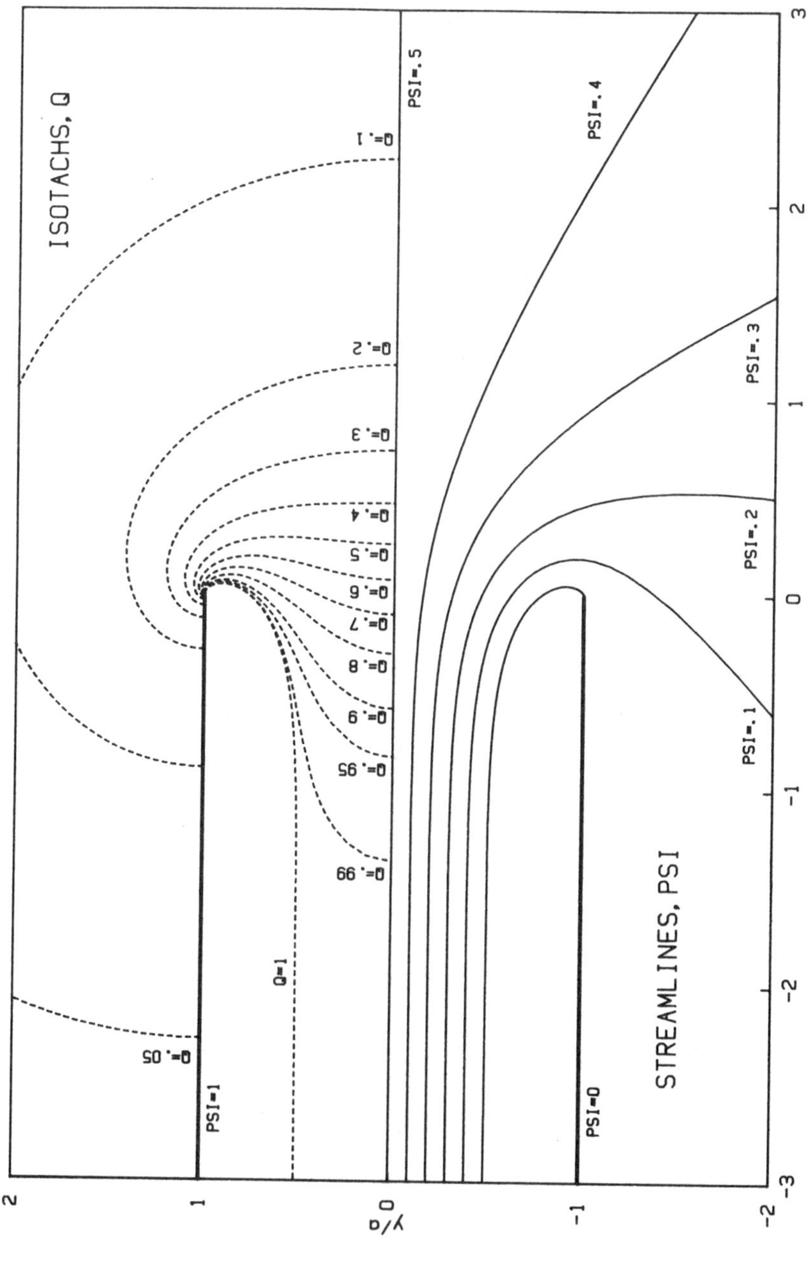

Figure 61 Borda's mouthpiece.

where the t plane is linked to the physical plane through the transformation

$$z = \frac{-2\ell}{\pi + 4}\left(1/t + \frac{(t^2 - 1)^{1/2}}{2t^2} + \frac{1}{2}\sin^{-1}\frac{1}{t}\right) \tag{159}$$

At the edge of the plate, the flow separates to form a free streamline along which both the speed and the pressure are constant. For simplicity, the speed s along the free streamline is taken to be s = 1. When the complex variable $t = \rho \exp(i\nu)$, the streamline equation may be written from (158) as

$$\text{PSI} = \frac{\sin 2\nu}{\rho^2(\pi + 4)} \tag{160}$$

where PSI = $\psi/s\ell$. Streamlines in the t plane are transformed to the physical plane through (159) and plotted on the left half of Figure 62.

The complex velocity is also given as

$$-u + iv = -\frac{1}{\zeta} = t - (t^2 - 1)^{1/2} \tag{161}$$

Thus $Q^2 = 1/\zeta\bar{\zeta}$. Isotachs are circles in the ζ plane which are transformed to the physical plane through (161) and (159) as plotted on the right half of Figure 62. The speed remains finite at the sharp edge in contrast to the flow of Figure 44 which does not separate and form a free streamline. Far downstream of the plate in Figure 62, the streamlines continue to separate without limit.

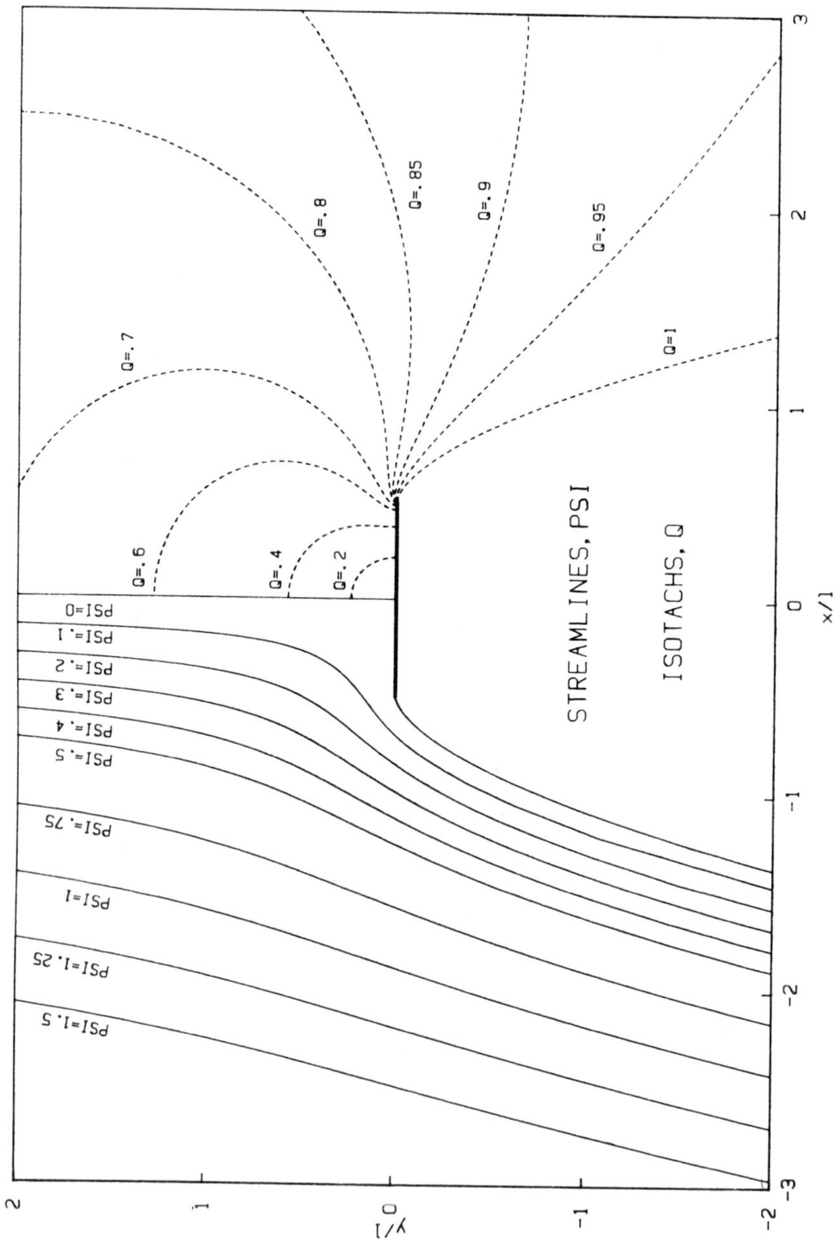

Figure 62 Separated flow past a plate.

REFERENCES

Currie, I. G., 1974. *Fundamental Mechanics of Fluids*, McGraw-Hill, New York.

Eskinazi, S., 1967. *Vector Mechanics of Fluids and Magnetofluids*, Academic Press, New York.

Feynman, R. P., R. B. Leighton and M. Sands, 1964. *The Feynman Lectures on Physics*, Volume II, Addison-Wesley, Reading, Massachusetts.

Fuller, A. J. B., 1973. *Engineering Field Theory*, Pergamon Press, Elmsford, New York.

James, M. L., G. M. Smith and J. C. Wolford, 1967. *Applied Numerical Methods for Digital Computation with FORTRAN*, International Textbook Co., Scranton, Pennsylvania.

Kennard, E. H., 1967. Irrotational Flow of Frictionless Fluids, Mostly of Invariable Density, David Taylor Model Basin Report 2299, NTIS, AD 653463, Springfield, Virginia.

Lamb, H., 1932. *Hydrodynamics*, Dover, New York.

Michelson, I., 1970. *The Science of Fluids*, Van Nostrand Reinhold, New York.

Milne-Thomson, L. M., 1968, 4th ed. *Theoretical Hydrodynamics*, Macmillan, New York.

Milne-Thomson, L. M., 1966, 4th ed. *Theoretical Aerodynamics*, Macmillan, London.

Streeter, V. L., 1948. *Fluid Dynamics*, McGraw-Hill, New York.

Taylor, G. I., 1963. In "The Scientific Papers of G. I. Taylor," Vol. 3, pp. 383-386, Cambridge University Press, Cambridge.

Van Dyke, M., 1982. *An Album of Fluid Motion*, The Parabolic Press, Stanford, California.

Weber, E., 1950. *Electromagnetic Theory, Static Fields and Their Mapping*, John Wiley and Sons, Inc., New York.

2
Three Dimensional Flows

2.1 INTRODUCTION

For three dimensional axisymmetric flows, a stream function ψ exists which is called the Stokes stream function. The difference between it and the two dimensional stream function used in Chapter 1 is that the volume flow between any two stream surfaces along which the Stokes stream function ψ is a constant is $2\pi\Delta\psi$. All the problems discussed in this section will be axisymmetric flows.

Flows referenced from Currie (1974) utilize an axisymmetric coordinate system (r, θ, ω). Because the flow is axisymmetric, there is no variation in the flow as ω varies from 0 to 2π. Thus, this system reduces to plane polar coordinates.

Flows referenced from Kennard (1967) utilize a similar axisymmetric coordinate system $(x, \tilde{\omega})$ in which the x axis is the reference axis and the variable $\tilde{\omega} = (y^2 + z^2)^{1/2}$. Since there is no variation of flow in the z direction, this system also reduces to a plane polar coordinate system in which

$x = r \cos \theta$, $y = r \sin \theta$, and $r = (x^2 + \tilde{\omega}^2)^{1/2}$. All the plots presented in Chapter 2 utilize the coordinates (x,y) to replace $(x, \tilde{\omega})$ because the character $\tilde{\omega}$ was not readily available.

2.2 THE POINT DIPOLE

Kennard (1967), page 302, gives the stream function for a point dipole of strength μ at the origin.

$$\psi = -\mu \frac{\tilde{\omega}^2}{r^3} \tag{162}$$

where $\tilde{\omega} = r \sin \theta$. The dimensionless stream function is PSI = $\psi a/\mu$ and streamlines are plotted directly from (162) on the upper half of Figure 63. The velocity components are also given as

$$q_x = \frac{\mu}{r^3}\left(\frac{3x^2}{r^2} - 1\right)$$

$$q_{\tilde{\omega}} = 3\mu \frac{x\tilde{\omega}}{r^5} \tag{163}$$

The dimensionless isotachs are defined to be lines of constant Q where

$$Q = \frac{a^3}{\mu} (q_x{}^2 + q_{\tilde{\omega}}{}^2)^{1/2} \tag{164}$$

Isotachs are plotted on the lower half of Figure 63. The flow in Figure 63 should be compared to the two-dimensional dipole plotted on Figure 6.

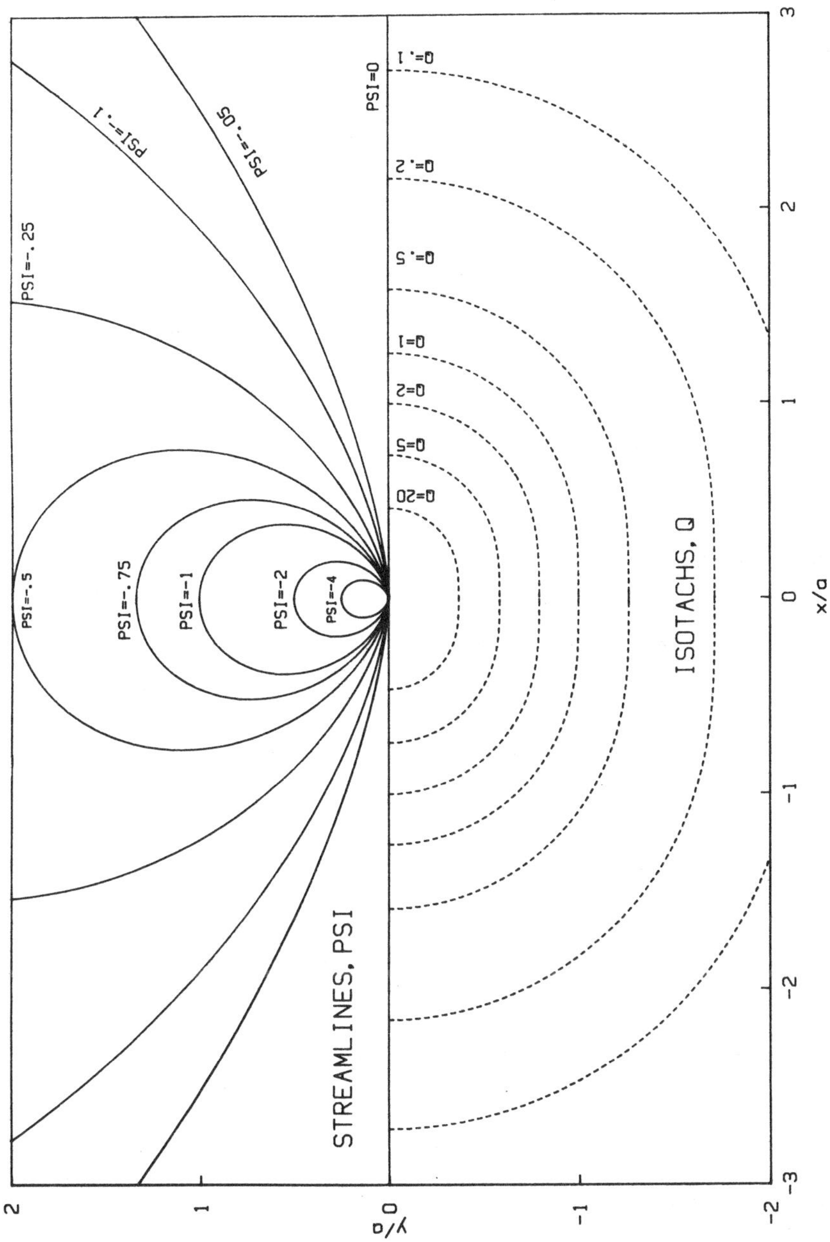

Figure 63 The point dipole.

2.3 FLOW OVER A SPHERE

Currie (1974), page 157, gives the stream function for the sphere of radius a in a uniform flow as

$$\psi(r,\ \theta) = \frac{1}{2}U\left(r^2 - \frac{a^3}{r}\right)\sin^2\theta \tag{165}$$

When the dimensionless stream function is defined PSI = ψ/Ua^2, equation (165) may be written

$$PSI = \frac{1}{2}\left(\left(\frac{r}{a}\right)^2 - \left(\frac{a}{r}\right)\right)\sin^2\theta \tag{166}$$

Equation (166) was solved for r = f(PSI, θ) and the streamlines are plotted on Figure 64.

The velocity components are calculated from (165) as

$$u_r = U\cos\theta\left(1 - \left(\frac{a}{r}\right)^3\right)$$

$$u_\theta = -U\sin\theta\left(1 + \frac{1}{2}\left(\frac{a}{r}\right)^3\right) \tag{167}$$

Thus, the isotach equation may be written as

$$Q^2 = \left(u_r^{\ 2} + u_\theta^{\ 2}\right)/U^2 \tag{168}$$

Equation (168) was solved for r = f(θ, Q) using a Newton-Raphson technique and isotachs are plotted on the lower half of Figure 64. Figure 12 for the flow over a cylinder should

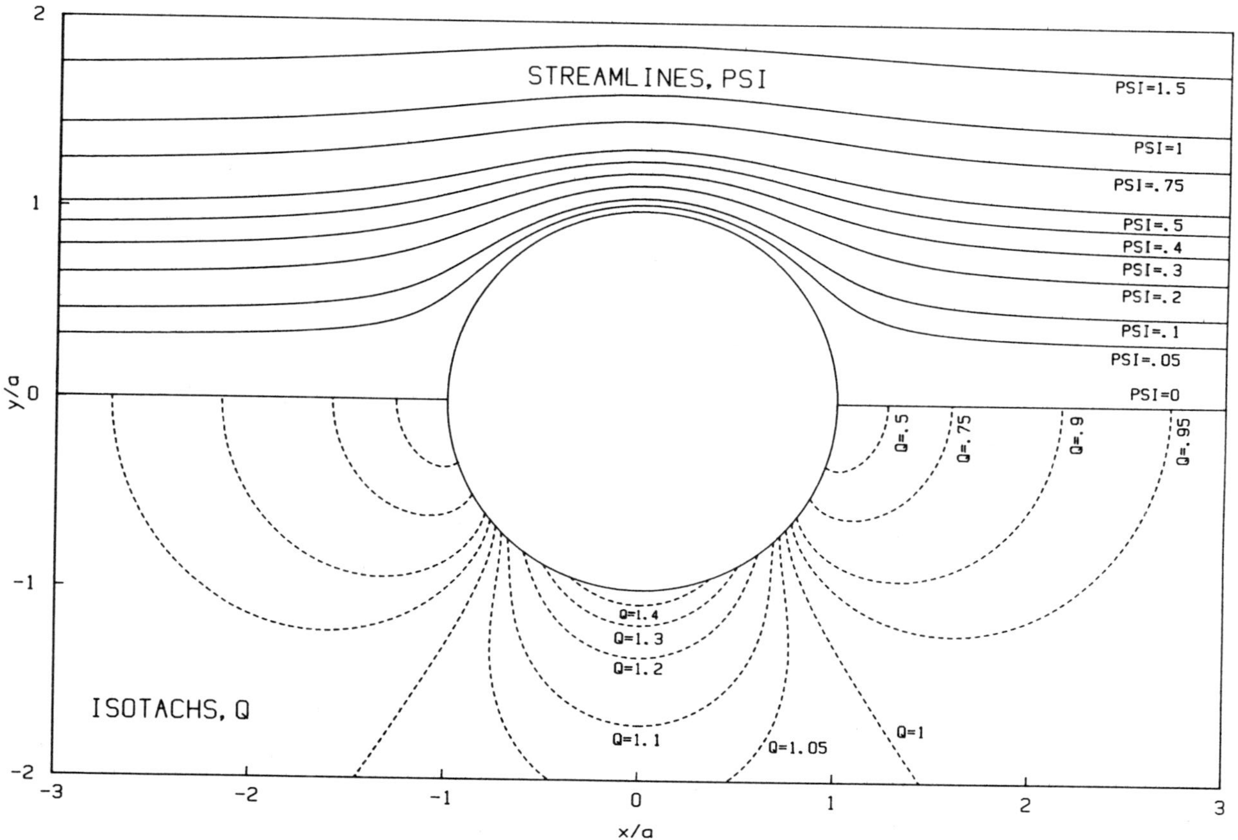

Figure 64 The sphere in a uniform flow.

be compared to Figure 64. The maximum speed on the sphere is $Q = 1.5$. The sphere perturbs the uniform flow less than the two dimensional cylinder.

2.4 THE TRANSLATING SPHERE

Kennard (1967), page 319, gives the stream function for a sphere of radius a translating at speed U into a fluid at rest.

$$\psi = -\frac{a^3U}{2r}\sin^2\theta \tag{169}$$

The dimensionless stream function is $PSI = \psi/Ua^2$ and streamlines are plotted on the upper half of Figure 65. The stream function (169) is identical to the stream function for a point dipole, equation (162), when the dipole strength is $\mu = a^3U/2$.

The speed at any point in the flow is also given as

$$Q^2 = \left(\frac{a}{r}\right)^6\left(1 - \frac{3}{4}\sin^2\theta\right) \tag{170}$$

where $Q = q/U$. Isotachs are plotted on the lower half of Figure 65. The minimum speed on the sphere is $Q = 0.5$ at $\theta = \pi/2$. Figure 13 for the translating cylinder should be compared to Figure 65. Notice that every point on the cylinder moves at $Q = 1$ while only the points $r = a, \theta = 0, \pi$ have the speed $Q = 1$ on the sphere.

2.5 THE RANKINE HALFBODY

Currie (1974), page 155, gives the stream function for a point source of strength m in a uniform flow.

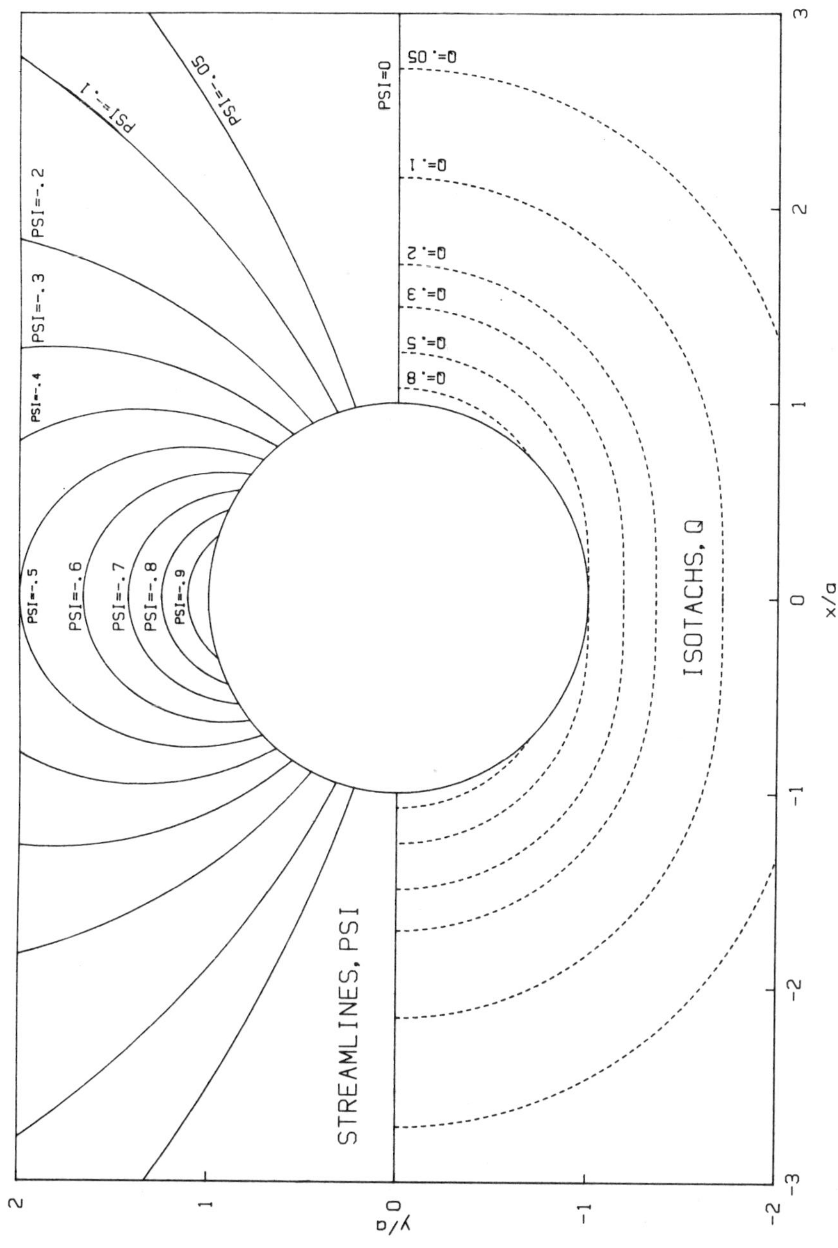

Figure 65 The translating sphere.

$$\psi(r,\theta) = \frac{1}{2} U r^2 \sin^2\theta - \frac{m}{4\pi}(1 + \cos\theta) \tag{171}$$

This superposition forms an open half-body defined by $\psi = 0$. The nose radius of the half body is $r_0 = (m/4\pi U)^{1/2}$ at $\theta = \pi$. Dimensionless streamlines are defined to be PSI = $\psi/Ur_0{}^2$ and the streamline equation is written

$$\frac{r}{r_0} = \left(\frac{1}{\sin^2(\theta/2)} + \frac{2\,\mathrm{PSI}}{\sin^2\theta}\right) \tag{172}$$

Streamlines are plotted on the upper half of Figure 66.

The velocity components calculated from (171) are

$$\frac{u_r}{U} = \cos\theta + \left(\frac{r_0}{r}\right)^2$$

$$\frac{u_\theta}{U} = -\sin\theta \tag{173}$$

Thus, the isotach equation is written

$$Q^2 = 1 + \left(\frac{r_0}{r}\right)^4 + 2\cos\theta\left(\frac{r_0}{r}\right)^2 \tag{174}$$

Equation (174) was solved using the quadratic formula for $r = f(\theta, Q)$ and isotachs are plotted on the lower half of Figure 66. The flow stagnates at the nose, accelerates to $Q = 1$ at $\theta = 109.5°$, further accelerates to a maximum $Q = 1.155$ at $\theta = 71°$, and the decelerates to $Q = 1$ far downstream on the body. This flow should be compared to its two dimensional counterpart plotted on Figure 9.

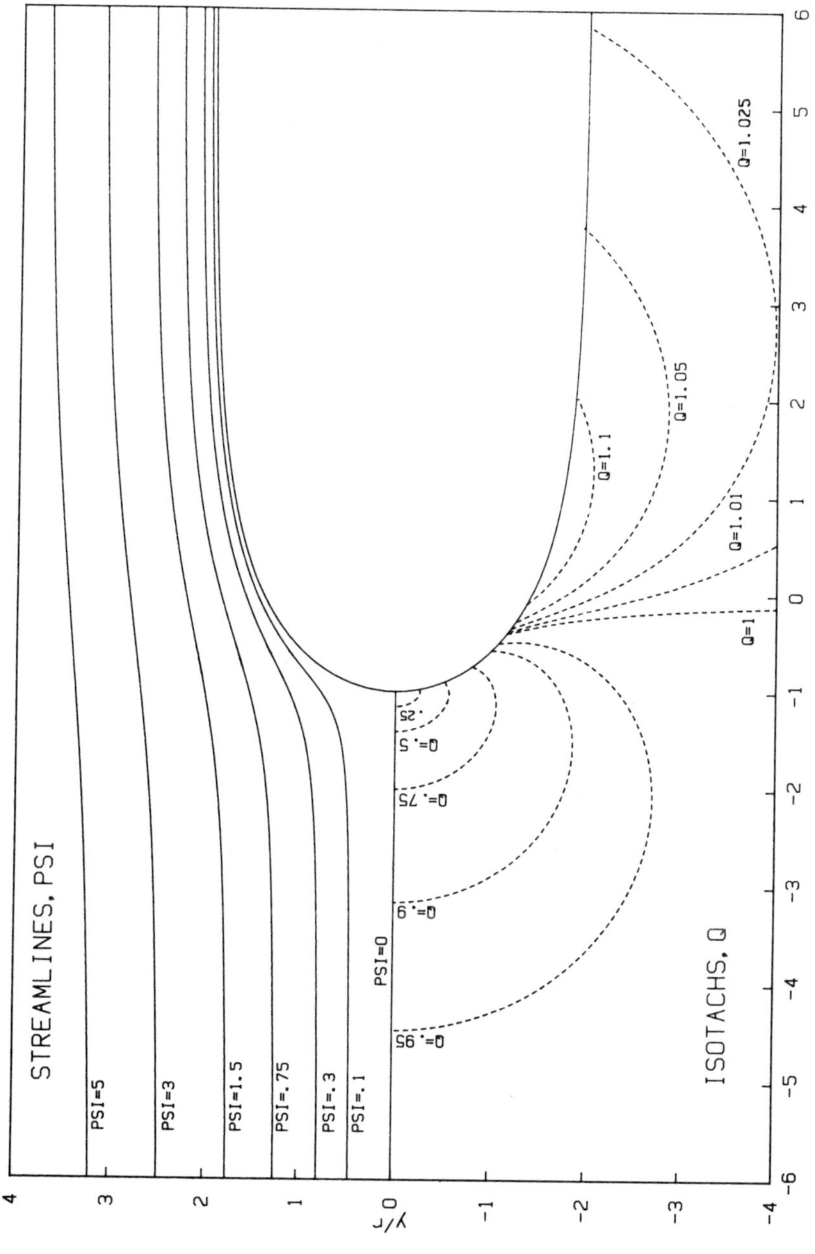

Figure 66 The Rankine halfbody.

2.6 THE RANKINE SOLID

Kennard (1967), page 308, gives the stream function ψ for a point source at x = a and an equal point sink at x = -a in a uniform stream of speed U as

$$\psi = \frac{1}{2} U \left(\tilde{\omega}^2 + b^2(\cos \theta_1 - \cos \theta_2)\right) \tag{175}$$

where the subscripts 1,2 refer to bipolar coordinates and b is a constant proportional to the strength of the source. This superposition forms a closed body shape given by

$$\left(\frac{\tilde{\omega}}{a}\right)^2 = \left(\frac{b}{a}\right)^2 (\cos \theta_2 - \cos \theta_1) \tag{176}$$

The dimensionless stream function is PSI = ψ/Ua^2 and streamlines are plotted from (175) on the upper half of Figure 67 for $(b/a)^2 = 0.7$.

The velocity components are given as

$$\frac{q_x}{U} = \left(-1 + \frac{\Lambda}{2}\left(\frac{x-1}{r_1^3} - \frac{x+1}{r_2^3}\right)\right)$$

$$\frac{q_{\tilde{\omega}}}{U} = \frac{1}{2} \Lambda \tilde{\omega} \left(\frac{1}{r_1^3} - \frac{1}{r_2^3}\right) \tag{177}$$

where $\Lambda = (b/a)^2$. Thus, isotachs are lines of constant

$$Q^2 = \left(\frac{q_x}{U}\right)^2 + \left(\frac{q_{\tilde{\omega}}}{U}\right)^2 \tag{178}$$

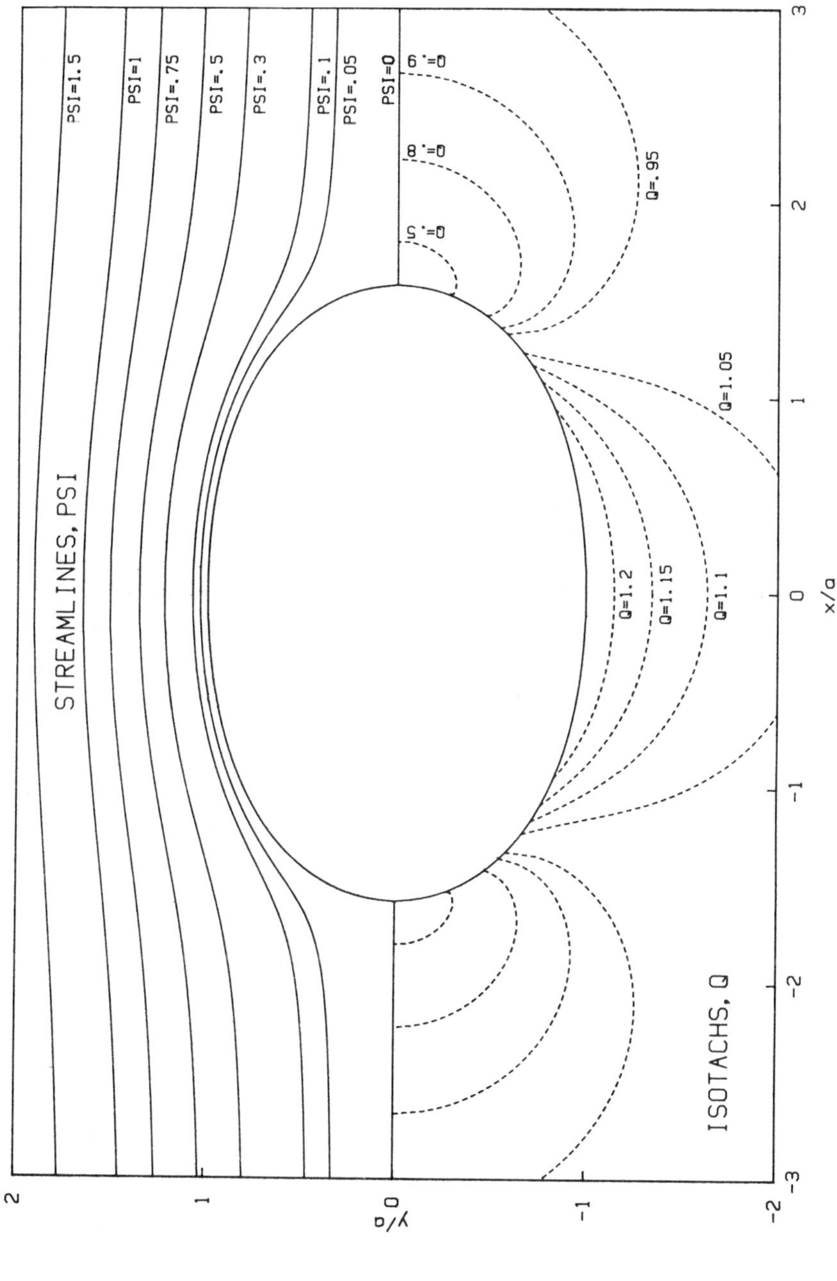

Figure 67 The Rankine solid.

Equation (178) was solved numerically using a Newton-Raphson technique for $r = f(\theta, Q)$ and isotachs are plotted on Figure 67. The flow stagnates at the nose and accelerates to a maximum of $Q = 1.25$.

2.7 A LINE OF POINT SOURCES

Kennard (1967), page 312, gives the stream function for a distribution of point sources along the x axis $a \leqslant x \leq b$ as

$$\psi = \alpha(r_a - r_b) \tag{179}$$

where the total volume flow from the distribution is $4\pi\alpha(b - a)$ and r_a and r_b denote the distances from the end of the line of sources to a field point. The dimensionless stream function is $PSI = \psi/\alpha b$ and streamlines are generated from (179) using a Newton-Raphson technique as plotted on the upper half of Figure 68 for the case of $a = -1$, $b = 1$.

The velocity components are also given as

$$q_x = \alpha(1/r_a - 1/r_b)$$

$$q_{\tilde{\omega}} = \frac{\alpha}{\tilde{\omega}}\left(\frac{x - a}{r_a} - \frac{x - b}{r_b}\right) \tag{180}$$

Isotachs are lines of constant Q where

$$Q = \frac{b}{\alpha}\left(q_x{}^2 + q_{\tilde{\omega}}{}^2\right)^{1/2} \tag{181}$$

Equation (181) was solved for $y = f(x, Q)$ and isotachs are plotted on the lower half of Figure 68. When the two dimensional distribution of line sources, plotted on Figure 41, is

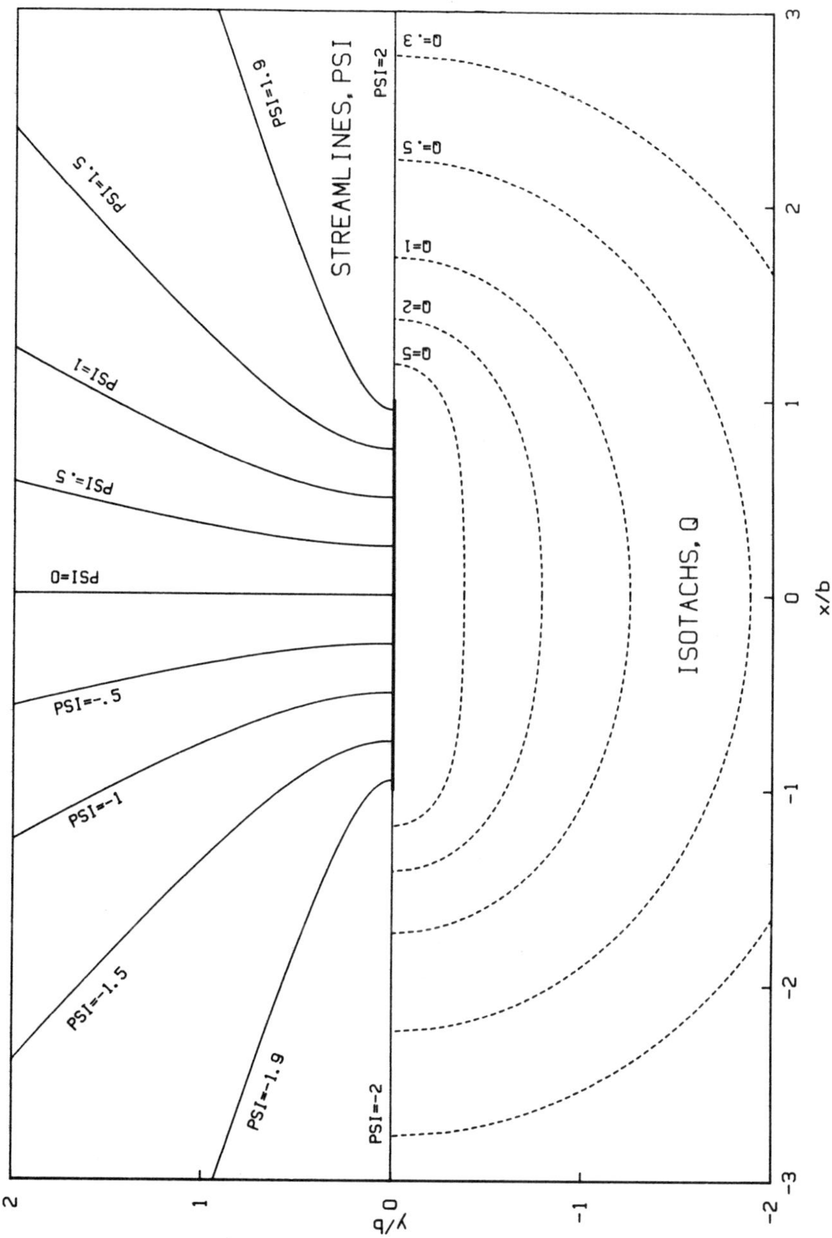

Figure 68 A line of point sources.

compared to this problem, it is seen that the speed is infinite only at the edges of the two dimensional distribution whereas the speed is infinite everywhere on the three dimensional distribution.

2.8 THE AIRSHIP

Kennard (1967), page 314, gives the stream function for a point source of strength $4\pi A$ at the origin plus a distributed point sink of equal total strength $-4\pi\alpha a$ along $-a \leqslant x \leqslant 0$ in a uniform flow from right to left.

$$\psi = U\left(\frac{\tilde{\omega}^2}{2} + b^2\frac{(r_1 - r)^2 - a^2}{2ar}\right) \tag{182}$$

where $b^2 = A/U$ and $r = (x^2 + \tilde{\omega}^2)^{1/2}$, $r_1 = ((x + a)^2 + \tilde{\omega}^2)^{1/2}$. This combination gives rise to a closed bodyshape of an airship in a flow. The equation for the body shape is

$$\frac{\tilde{\omega}}{a}^2 = \Lambda\left(\frac{1 - (r_1/a - r/a)^2}{r/a}\right) \tag{183}$$

where the parameter $\Lambda = (b/a)^2$ controls the thickness ratio of the airship. The dimensionless stream function is PSI = ψ/Ua^2. Streamlines were constructed by solving (182) using a Newton-Raphson technique for y = f(x,PSI) and together with the body shape are plotted on Figure 69 for $\Lambda = 0.048$.

The velocity components are given as

$$q_x = U(-1 + b^2(x/r^3 - (1/r - 1/r_1)/a))$$

$$q_{\tilde{\omega}} = U\left(\frac{\tilde{\omega}}{r^3} - \frac{1}{a\tilde{\omega}}\left(\frac{(x + a)}{r_1} - \frac{x}{r}\right)\right) \tag{184}$$

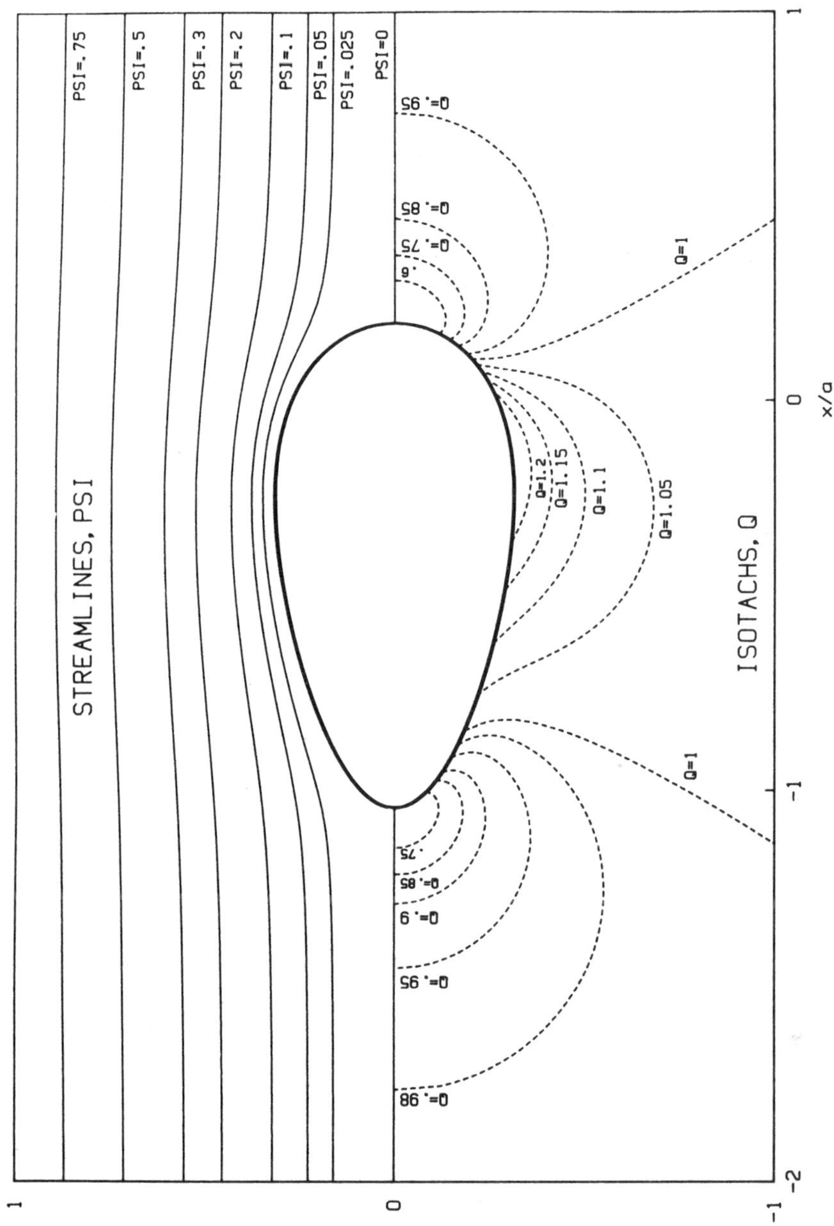

Figure 69 The airship.

Isotachs are lines of constant

$$Q = (q_x^{\ 2} + q_{\tilde{\omega}}^{\ 2})/U \tag{185}$$

Equation (185) was solved for y = f(x, Q) using a Newton-Raphson technique and the isotachs are plotted on the lower half of Figure 69. The flow stagnates at the nose of the airship, accelerates to a maximum of Q = 1.275 at the widest point and then decelerates to stagnation at the rear.

2.9 THE CIRCULAR APERTURE

Kennard (1967), page 369, gives the stream function for the flow through a circular aperture of width 2c

$$\psi = A\ c\ \mu \tag{186}$$

where the volume flow through the aperture is $2\pi cA$ and the speed at the origin is U = A/c. The parameters $x, \tilde{\omega}$ are related to μ, ζ through the oblate-spherical coordinate equations

$$x = c\ \mu\ \zeta$$

$$\tilde{\omega} = c(\zeta^2 + 1)^{1/2}(1 - \mu^2)^{1/2} \tag{187}$$

The dimensionless stream function PSI = ψ/Ac and the dimensionless streamline equation may be written from (186) and (187) as

$$\frac{\tilde{\omega}}{c} = \left((x/PSI)^2 + 1\right)^{1/2}\left(1 - PSI^2\right)^{1/2} \tag{188}$$

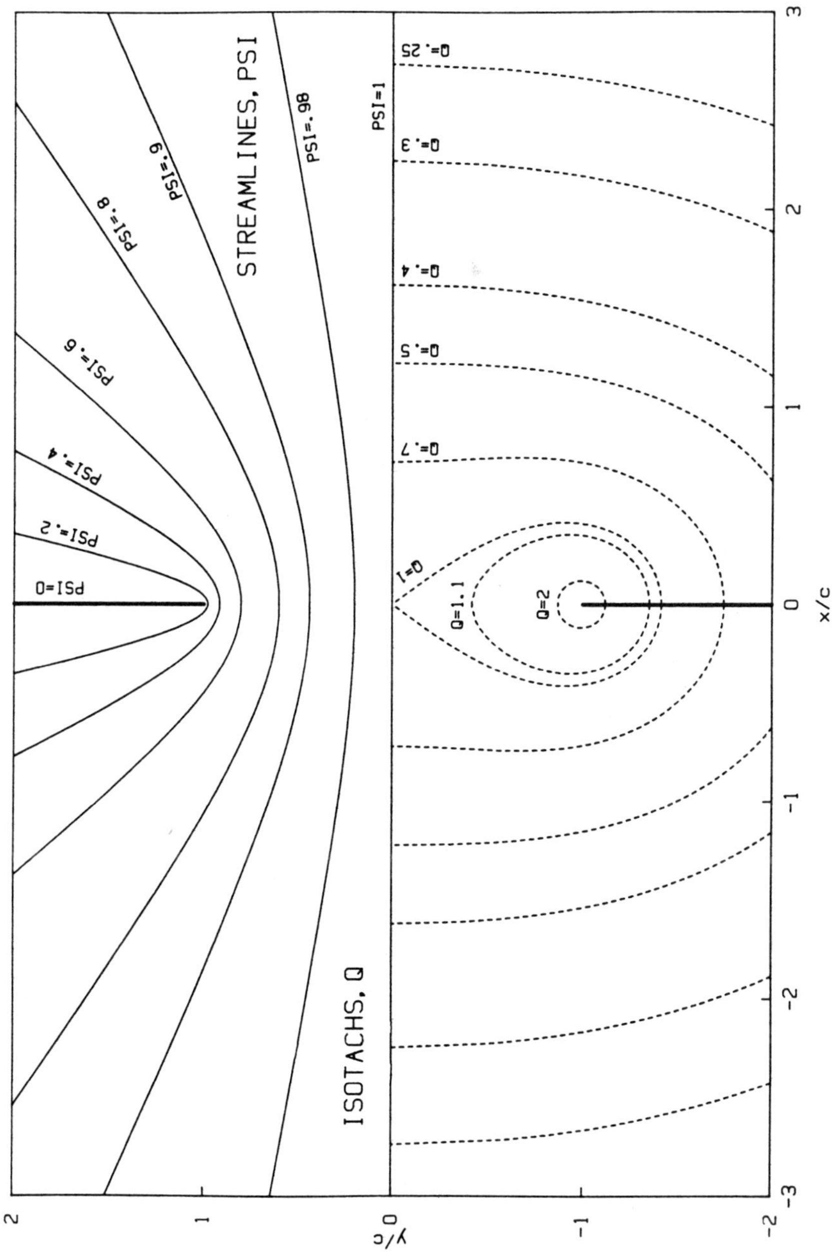

Figure 70 The circular aperture.

Equation (188) was solved directly and streamlines are plotted on the upper half of Figure 70.

The speed at any point in the flow is given as

$$Q = \left((\zeta^2 + 1)/(\zeta^2 + \mu^2) \right)^{1/2} \tag{189}$$

where $Q = q/U$. Equation (189) was solved for $\zeta = f(\mu, Q)$ and coordinates were transformed through (187) as plotted on the lower half of Figure 70. The speed goes to infinity at the sharp edge. Since any streamline can be a solid surface, Figure 70 represents a family of aperture flows. Figure 70 should be compared with its two dimensional counterpart plotted on Figure 24.

REFERENCES

Currie, I. G., 1974. *Fundamental Mechanics of Fluids*, McGraw-Hill, New York.

Kennard, E. H., 1967. Irrotational Flow of Frictionless Fluids, Mostly of Invariable Density, David Taylor Model Basin Report 2299, NTIS, AD653463, Springfield, Virginia.

3
Worked Examples

3.1 INTRODUCTION

The first step in making use of the graphical solutions presented herein is to have some way of interpolating between lines and adjusting to the variety of scales of the different plots. An excellent instrument for this task is the Gerber Variable Scale manufactured by the Gerber Scientific Instrument Company of Hartford, Connecticut. This clever instrument is designed around a precision spring which adjusts to any scale. The coils of the spring are marked to provide the calibration of an expandable scale. Direct measurements to any scale can be made by simply adjusting the spring length.

If the reader does not have access to a Gerber Scale, a very inexpensive but effective interpolator can be made from an acetate transparency of Figure 71. The use of this interpolator is almost self evident. The dashed lines are provided as alignment guides. The acetate can be cut in half to provide both a linear and a logarithmic interpolator.

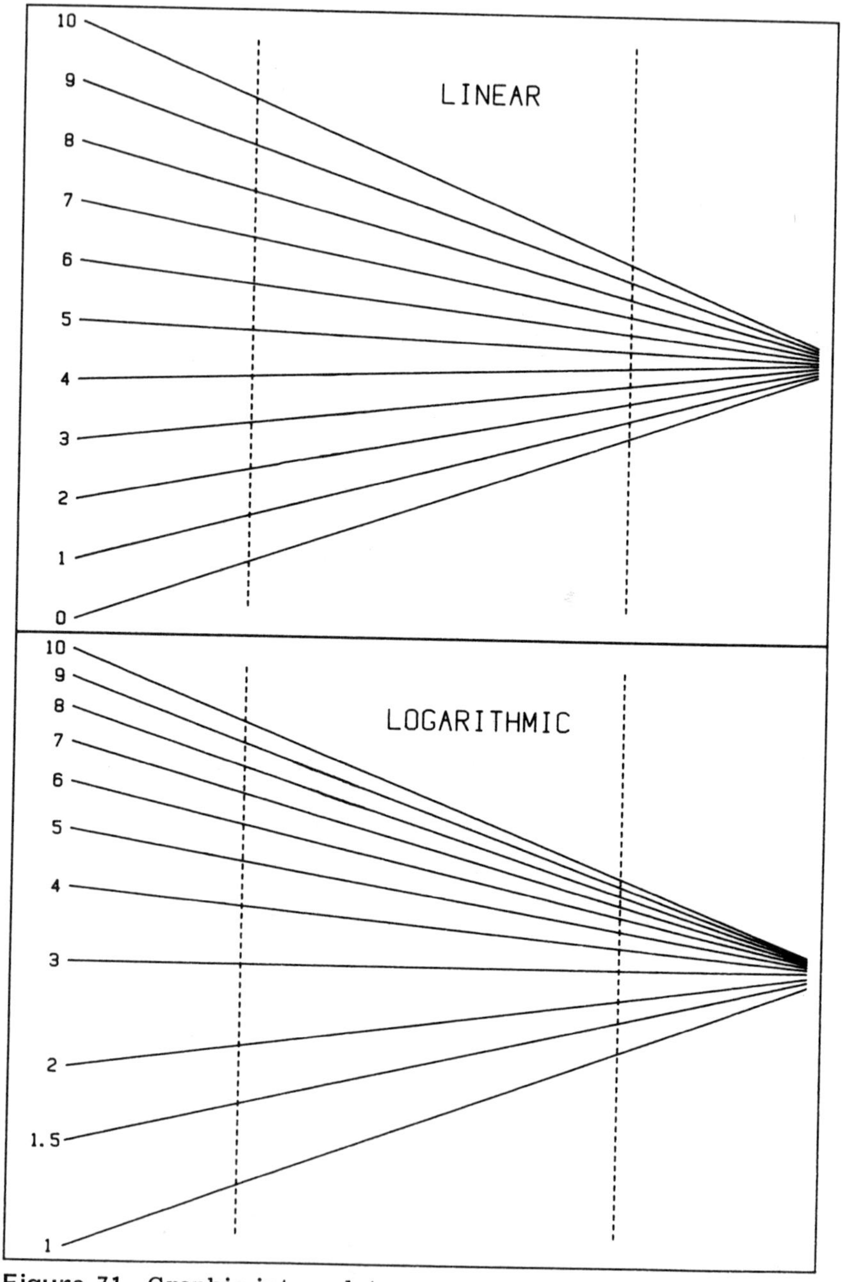

Figure 71 Graphic interpolator.

Worked Examples

The five examples which follow show only a few of the ways in which these graphical solutions can be utilized. The reader should bear in mind that these potential flows should only be used to make calculations of real flow situations outside boundary layers in regions of favorable pressure gradient where the flow has not separated.

3.2 FLOW PAST A PLATE

Compare the speed along a flat plate of height h situated transverse to a flow for the case of separated and nonseparated flow. The tabulated data below were scaled directly from Figures 44 and 62. The table clearly shows the effect of allowing the fluid to separate along the free streamlines. Several examples of real flow past a plate are depicted in the collection of Van Dyke (1982). The non-separated potential flow looks like real fluid flow only for very small Reynold's Numbers.

Height	y/h	0	0.2	0.4	0.6	0.8	1
Separated flow	Q	0	0.1	0.2	0.32	0.48	1
Nonseparated flow	Q	0	0.2	0.45	0.72	1.3	∞

3.3 EFFLUX FROM AN ORIFICE

Water is seen to be draining from a two dimensional orifice of the type shown in Figure 60 at an asymptotic speed of 0.5 m/s. What will be the pressure at selected points along the bottom tank wall, the PSI = 0 streamline? At x/a = 1, Q = 1 and, thus, the speed is 0.5 m/s and the pressure is atmospheric. At x/a = 2, Q = 0.22 and, thus, the speed is q =

0.2×0.5 m/s = 0.11 m/s. From Bernoulli's equation, the increase in pressure above atmospheric pressure is $\Delta p = 1/2\rho(q_1^2 - q_2^2)$. $\Delta P = 1/2\,(1000)\text{kg/m}^3[(0.5)^2 - (0.11)^2]\text{m}^2/\text{s}^2 = 118.95$ Pa. And finally at x/a = 3, Q = 0.125, q = 0.0625 m/s and $\Delta p = 123.05$ Pa.

3.4 NOZZLE DESIGN

A two dimensional venturi nozzle for an air duct of the vortex shaped nozzle design, Figure 37, is to be built. A gradual slope nozzle is desired, of approximately the shape of the PSI = 1 streamline. The upstream duct width at the start of the contraction is 10 cm. What is the duct width at the minimum area? From Figure 37, the PSI = 1 streamline contracts from y/a = 1.8 at x/a = 6 to y/a = 0.4 at x/a = 0. Thus, the minimum duct width is $10 \times 0.4/1.8 = 2.22$ cm. For an air flow of 3 m/s at the start of the contraction, x/a = 6, what pressure difference will be measured between a pressure tap at the minimum area and a pressure tap at x/a = 6? An approximate answer is obtained by assuming one dimensional flow. The speed at the minimum area is $q = 3 \times 1.8/04 = 13.5$ m/s. Thus, the pressure difference from Bernoulli's equation is

$$\Delta p = \frac{1}{2}\,\rho(q_2^{\,2} - q_1^{\,2}) = \frac{1}{2}\left(1.2\,\frac{\text{kg}}{\text{m}^3}\right)(13.5^2 - 3^2)\,\frac{\text{m}^2}{\text{s}^2}$$

$$= 103.95 \text{ Pa}$$

A more precise estimate for the speed along the duct wall at the minimum area is obtained from (93). At x/a = 0, y/a = 0.4; z = 0 + i(0.4)

$$Q^2 = \frac{[(0.4i)^2 + 5][(-0.4i)^2 + 5]}{[(0.4i)^2 + 1][(-0.4i)^2 + 1]} = 33.19$$

Thus, Q = 5.7619 and the speed on the wall at the minimum area is 17.286 m/s. This second estimate clearly reflects the non-uniform velocity distribution at the minimum area. The new pressure difference is then estimated to be Δp = 1/2 (1.2)kg/m^3[(17.286)^2 − 3^2]m^2/s^2 = 173.88 Pa.

3.5 CHECKING A COMPUTER CODE

A large computer code called MATHEW, in which the motion of the wind is predicted in a flow field from the measured wind input at several grid locations is to be checked against some potential flow models. For the case of flow over an unbounded step, Figure 47, the wind speed far upstream of the step is 3 m/s and the step is 1 m high. The wind speed and direction input at several grid locations, is to be checked against some 47, and listed in the table below. The angle theta from the

x-position x/h	1.5	0.5	1.25	-0.5
y-position y/h	0.5	1.0	1.75	1.5
Dimensionless speed Q	0.74	0.885	0.92	1.23
Angle θ	8°	32.2°	11°	6.5°
Speed (m/s) q	2.22	2.66	2.76	3.69

local slope of the streamlines is measured relative to the horizontal axis. Using the above data as input, the MATHEW code output can be checked at virtually any point in the flow field by interpolation from Figure 47. Difficulties are, of course, to be expected at the sharp corner where the potential flow predicts infinite speed.

3.6 FLOW OVER A BANK OF CYLINDERS

The flow over a bank of 3 cm diameter cylinders is to be approximated by an infinite row of doublets in a uniform stream as shown on Figure 22. The air speed far upstream of the cylinder bank is 1 m/s. What will be the maximum speed on top of one of the cylinders? Where should a pressure tab be located on one of the cylinders to measure the free stream pressure? What will be the center to center spacing of the cylinders?

The reader should appreciate that, except at very low Reynold's Numbers, the flow will separate on the downstream side of the the bank of cylinders. Thus, the solution presented on Figure 22 is only valid upstream of the bank of cylinders where the pressure gradient is favorable.

Assume that the cylinder semidiameter is the mean of the semidiameters of the cylindrical shape of Figure 22, $r/b = (0.8814 + 0.8603)/2 = 0.8709$. Thus, for a 3 cm diameter cylinder $b = 1.5/0.8709 = 1.7224$ and the cylinder separation is $a = \pi b = 5.4109$ cm. This scaling can be checked by directly measuring on Figure 22. The maximum dimensionless speed from the text is $Q = 2.74$; thus, the maximum air speed is on the cylinder $q = 2.74 \times 1$ m/s $= 2.74$ m/s. A measurement of the free stream pressure can be made on the cylinder at the point where the $Q = 1$ isotach intersects the cylinder. And from Figure 22, this intersection occurs at $x/b = -0.8$, $y/b = -0.35$. In terms of the 3 cm cylinder, these coordinates are $x = -1.3779$ cm, $y = 0.6028$ cm. Or measuring with a protractor directly from Figure 22, the $Q = 1$ streamline intersects the cylinder at an angle of 24.5° down from the stagnation streamline PSI = 0.

REFERENCE

Van Dyke, M., 1982. *An Album of Fluid Motion*, The Parabolic Press, Stanford, California.

Appendix: Sample Software

INTRODUCTION

The programs in the appendix are examples of the type of software developed for this book. The programs are written in BASIC using the Hewlett Packard Graphics Language and they were run on the HP 87 personal computer.

Line numbers up to approximately 250 are for the set up of the graphics. The main programs are from line numbers 250 to 999. Subroutines start at line 1000. The plotting routines common to all the programs are from line numbers 1000 to 1270 and 1300 to 1375. Line numbers 2000 through 2999 are reserved for complex transformations. Line numbers greater than 3000 are reserved for complex variable software. A typical Newton-Raphson inversion subroutine starts at line number 1580 of the first program listed.

The program titles in this Appendix are:

1. The Cylinder With Circulation.
2. Flow Over a Bounded Step.
3. Flow Over a Flat Plate With Circulation.
4. Symmetric Joukowski Airfoil, Alpha = 15.
5. Borda's Mouthpiece.

```
10 REM ****THE CYLINDER WITH CIRCULATION ****
20 REM ****SET UP AND LABEL
30 PLOTTER IS 705
40 INTEGER K,L,J,P,N
50 K=1 @ L=1
60 OPTION BASE 1
70 DIM V(2000),W(2000)
80 MAT V=( 0)
90 MAT W=( 0)
100 OUTPUT 705 ;"VS";2
110 GCLEAR
120 LIMIT 0,250,0,180
130 LOCATE 7.5,135,10,95
140 FRAME
150 SCALE -6,6,-4,4
160 FXD 0
170 CSIZE 3
180 LAXES -1,1,-6,-4,1,1,2
190 MOVE 0,-4.6 @ LORG 4
200 LABEL "x/a"
210 MOVE -6.4,0 @ DEG @ LDIR 90
220 LABEL "y/a"
230 MOVE 0,4.1 @ LDIR 0 @ CSIZE 6
240 LABEL "THE CYLINDER WITH CIRCULATION"
250 RAD
253 BETA=PI /12
260 REM *** DRAW THE CYLINDER ****
262 P=0 @ K=1 @ L=1
265 FOR TH=0 TO 2*PI +PI /180 STEP PI /180
270 X=COS (TH)
275 Y=SIN (TH)
280 GOSUB 1000
300 NEXT TH
305 GOSUB 1050
315 GOTO 320
320 REM **** DRAW PSI > 0 ****
322 DATA 1,-.25,-.5,-.75,-1,-1.5,-2,1.5,2,2.5,3,10
325 READ PSI
330 K=1 @ L=1 @ J=0
335 IF PSI=10 THEN 400
340 R=4
350 IF PSI>0 THEN TH=PI /2 ELSE TH=3*PI /2
351 IF PSI>0 THEN DTH=-(PI /90) ELSE DTH=PI /90
352 GOSUB 1400 @ J=J+1
355 GOSUB 1215
360 IF ABS (X)>6 THEN 390
361 IF ABS (Y)>4 THEN 390
385 TH=TH+DTH
387 GOTO 352
390 GOSUB 1130
395 GOTO 325
400 GOTO 415
415 REM **** PSI=0 ****
```

```
420 DATA -.2618,10
425 READ TH
427 IF TH=10 THEN 480
430 K=1 @ L=1 @ DR=.05 @ PSI=0
435 R=1
440 GOSUB 1215
445 R=R+DR
450 GOSUB 1455
455 GOSUB 1215
460 IF ABS (X)>6 THEN 470
465 GOTO 445
470 GOSUB 1130
475 GOTO 425
480 REM **** PSI>0 ****
484 DATA .25,.5,.75,10
486 READ PSI
488 K=1 @ L=1 @ J=0
490 IF PSI=10 THEN 525
492 TH=PI /2
494 DTH=PI /180 @ R=4 @ DR=.05
496 TH=TH-DTH
498 GOSUB 1400 @ GOSUB 1215
500 GOSUB 1000 @ J=J+1 @ DISP J
502 IF R>1.45 THEN 504 ELSE 496
504 R=R+DR
506 GOSUB 1455 @ GOSUB 1215 @ GOSUB 1000
508 J=J+1 @ DISP J
510 IF ABS (X)>6 THEN 512 ELSE 504
512 GOSUB 1130 @ GOTO 486
525 REM **** Q<>1 ****
526 DATA 1.1,4.7124,5.3407,1.25,4.7124,5.1967
527 DATA .95,4.7124,5.4585,1,4.7124,5.4192
528 DATA 1.05,6.5537,7.854,1.1,6.5798,7.854,1.25,6.6584,7.854
529 DATA 1.5,6.7936,7.854,2,7.1165,7.854,10,10,10
531 LINE TYPE 4,1
533 READ Q,TH1,TH2
535 K=1 @ L=1 @ P=P+1
537 IF Q=10 THEN 559
539 DTH=(TH2-TH1)/100
540 IF P<5 THEN TH=TH2 ELSE TH=TH1
541 RHO=1
543 GOSUB 1250
545 FOR J=1 TO 100
547 IF P<5 THEN TH=TH-DTH ELSE TH=TH+DTH
549 GOSUB 1580 @ DISP J
551 GOSUB 1250
553 NEXT J
555 GOSUB 1050
557 GOTO 533
559 DATA .5,5.7465,6.2745,10,10,10
561 READ Q,TH1,TH2
563 K=1 @ L=1
565 IF Q=10 THEN 605
567 TH=(TH1+TH2)/2 @ TH3=TH
```

```
568 IF Q=.5 THEN RHO=2 ELSE RHO=3
569 GOSUB 1580 @ RHOM=RHO @ PRINT RHOM
571 RHOB=1 @ DRHO=(RHOM-RHOB)/100
573 RHO=RHOB @ TH=TH1 @ GOSUB 1250
575 FOR J=1 TO 99
577 RHO=RHO+DRHO @ GOSUB 1650 @ GOSUB 1250 @ DISP J
579 IF ABS (Y)>4 THEN 583
581 NEXT J
582 RHO=RHOM @ TH=TH3 @ GOSUB 1250
583 GOSUB 1050 @ K=1 @ L=1
585 RHOB=1 @ DRHO=(RHOM-RHOB)/100
587 RHO=RHOB @ TH=TH2 @ GOSUB 1250
589 FOR J=1 TO 99
591 RHO=RHO+DRHO @ GOSUB 1650 @ GOSUB 1250 @ DISP J
593 IF ABS (Y)>4 THEN 597
595 NEXT J
596 RHO=RHOM @ TH=TH3 @ GOSUB 1250
597 GOSUB 1050
599 GOTO 561
605 REM **** Q=1 ****
607 Q=1
626 K=1 @ L=1 @ J=0
627 TH=6.5275
629 RHO=1 @ DRHO=.05
631 GOSUB 1250
633 RHO=RHO+DRHO
635 GOSUB 1650 @ J=J+1 @ DISP J
637 GOSUB 1250
639 IF ABS (X)>6 THEN 643
640 IF ABS (Y)>4 THEN 643
641 GOTO 633
643 GOSUB 1050
645 GOTO 730
730 REM **** Q<1 ****
731 DATA .9,5.4977,6.4751,.035
732 DATA .85,5.5283,6.448,.028
735 DATA .75,5.5937,6.4,.02,10,10,10,10
740 READ Q,TH1,TH2,DRHO
745 IF Q=10 THEN 865
750 K=1 @ L=1
755 TH=TH1 @ RHO=1 @ GOSUB 1250
765 FOR J=1 TO 50
770 RHO=RHO+DRHO
775 GOSUB 1650 @ DISP J
780 GOSUB 1250
785 NEXT J
790 TH3=TH @ RHO3=RHO @ GOSUB 1050 @ K=1 @ L=1 @ RHO=1
791 TH=TH2 @ GOSUB 1250
795 FOR J=1 TO 50
800 RHO=RHO+DRHO
805 GOSUB 1650 @ DISP J
810 GOSUB 1250
815 NEXT J
817 GOSUB 1050 @ K=1 @ L=1
```

```
820 RHO4=RHO @ TH4=TH
825 DTH=(TH4-TH3)/50
827 TH=TH3 @ RHO=RHO3 @ GOSUB 1250
830 FOR J=1 TO 50
835 TH=TH+DTH
840 GOSUB 1580 @ DISP J
845 GOSUB 1250
850 NEXT J
855 GOSUB 1050
860 GOTO 740
865 GOTO 867
867 REM **** FINAL LABELING STREAMLINES ****
868 CSIZE 4 @ LORG 4 @ DEG @ LDIR 0
869 MOVE -2,3.6 @ LABEL "BETA=15 DEGREES"
870 LORG 1 @ MOVE -5.5,3 @ LABEL "STREAMLINES,PSI"
871 X=-4.5 @ CSIZE 2.5 @ LDIR 5.5 @ LORG 7
872 DATA -2,-2.88,-1.5,-2.36,-1,-1.83,-.75,-1.56,-.5,-1.29
873 DATA .25,-.49,.5,-.21,.75,.04,1,.3,1.5,.84,2,1.34
874 DATA -.25,-1.01,0,-.75,2.5,1.85,3,2.35
875 FOR J=1 TO 15
876 READ PSI,Y@ MOVE X,Y
877 LABEL USING 878 ; "PSI=",PSI
878 IMAGE K,K
879 NEXT J
880 GOTO 882
882 REM **** FINAL LABELING ISOTACHS ****
883 CSIZE 4 @ LORG 4 @ DEG @ LDIR 0
885 MOVE 4,-3.5 @ LABEL "ISOTACHS,Q"
886 CSIZE 2.5 @ X=.04
887 DATA 1.34,2,2.07,1.5,3.35,1.25,-2.63,.95,-2,1
889 FOR J=1 TO 5
890 READ Y,Q
891 IF J<4 THEN LORG 1 ELSE LORG 3
892 MOVE X,Y @ LABEL USING 893 ; "Q=",Q
893 IMAGE K,K
894 NEXT J
895 LORG 3 @ CSIZE 2
896 MOVE .01,-1.27 @ LABEL "1.25" @ MOVE .01,-1.55
897 LABEL "1.1" @ CSIZE 2.5 @ LDIR 0 @ LORG 3
898 DATA 1.47,-1.18,.75
899 DATA 3,3,1.1,4.38,2.9,1.05,4.6,1.15,1,1.1,-.69,.5
900 DATA 1.8,-1.8,.85,2.3,-2.75,.9
901 FOR J=1 TO 7
902 READ X,Y,Q
903 MOVE X,Y
904 LABEL USING 893 ; "Q=",Q
905 NEXT J
906 END
1000 REM **** STORES X,Y, AND COUNTS
1010 V(K)=X @ W(K)=Y
1020 L=K
1030 K=K+1
1040 RETURN
1050 REM **** PLOTS POINTS WITHOUT REFLECTION ****
```

```
1060 K=1
1070 MOVE V(K),W(K)
1080 FOR K=1 TO L
1090 PLOT V(K),W(K),-1
1100 NEXT K
1110 PEN UP
1120 RETURN
1130 REM **** PLOTS POINTS X AXIS REFLECTED ****
1140 K=1
1150 MOVE -V(K),W(K)
1160 FOR K=1 TO L
1170 PLOT -V(K),W(K),-1
1180 NEXT K
1190 PEN UP
1200 RETURN
1210 END
1215 REM **** CART TO POLAR & STORE ****
1220 X=R*COS (TH) @ Y=R*SIN (TH)
1225 GOSUB 1000
1230 RETURN
1250 REM **** CARTESIAN TO POLAR & STORE ****
1255 X=RHO*COS (TH)
1260 Y=RHO*SIN (TH)
1265 GOSUB 1000
1270 RETURN
1275 REM **** DECREMENT & STORE RHO & NU ON A SL ****
1280 FOR J=1 TO 1499
1285 R=R-DR
1290 GOSUB 1455
1295 GOSUB 1215
1296 NEXT J
1297 RETURN
1300 REM **** PLOTS POINTS WITH BOTH X & Y REFLECTED ****
1305 K=1
1310 MOVE -V(K),-W(K)
1315 FOR K=1 TO L
1320 PLOT -V(K),-W(K),-1
1325 NEXT K
1330 PEN UP
1335 RETURN
1340 REM **** PLOTS POINTS WITH Y REFLECTED ****
1345 K=1
1350 MOVE V(K),-W(K)
1355 FOR K=1 TO L
1360 PLOT V(K),-W(K),-1
1365 NEXT K
1370 PEN UP
1375 RETURN
1400 REM **** RETURNS R FOR TH & PSI ON A SL ****
1402 A=(R^2-1)/R
1405 B=A*SIN (TH)
1410 C=2*SIN (BETA)*LOG (R)
1420 FR=B+C-PSI
1425 FPR=SIN (TH)*((R^2+1)/R^2)+2*SIN (BETA)/R
```

```
1430 RT=FR/FPR
1435 IF ABS (RT)<.0001 THEN 1450
1440 R=R-RT
1445 GOTO 1402
1450 RETURN
1455 REM **** RETURNS TH FOR R&PSI ON A SL ****
1465 A=PSI-2*SIN (BETA)*LOG (R)
1470 B=R/(R^2-1)
1475 C=A*B
1480 TH=ASN (C)
1485 RETURN
1580 REM **** RETURNS RHO FOR Q & NU --N & R ****
1585 DRHO=.05 @ E=.0001
1590 GOSUB 2100
1595 F1=FRHO
1600 RHO=RHO+DRHO
1605 GOSUB 2100
1610 F2=FRHO
1615 F3=(F2-F1)/DRHO
1617 IF F3=0 THEN 1640
1620 RT=F1/F3
1625 IF ABS (RT)<E THEN 1640
1630 RHO=RHO-DRHO-RT
1635 GOTO 1590
1640 RHO=RHO-DRHO
1645 RETURN
1650 REM **** RETURNS TH FOR Q&RHO - NR ****
1655 DTH=.015 @ E=.0001
1660 GOSUB 2100
1665 F1=FRHO
1670 TH=TH+DTH
1675 GOSUB 2100
1680 F2=FRHO
1685 F3=(F2-F1)/DTH
1690 IF F3=0 THEN 1715
1695 RT=F1/F3
1700 IF ABS (RT)<.0001 THEN 1715
1705 TH=TH-DTH-RT
1710 GOTO 1660
1715 TH=TH-DTH
1720 RETURN
2000 REM **** ZETA AS RHO&NU TO Z AS R&I - JOUKOWSKI ****
2005 R8=RHO*COS (NU)
2010 I8=RHO*SIN (NU)
2015 N=-1
2020 R1=R8 @ I1=I8
2025 GOSUB 3050
2030 R=R+R8
2035 I=I+I8
2040 RETURN
2100 REM **** RETURNS Q FOR NU ON RHO=1 ****
2105 A=2*SIN (BETA)
2110 R10=RHO*COS (TH) @ I10=RHO*SIN (TH)
2115 R1=R10 @ I1=I10 @ N=-1 @ GOSUB 3050
```

```
2120 R9=R @ I9=I
2125 R1=R10 @ I1=I10 @ N=-2 @ GOSUB 3050
2130 R8=R @ I8=I
2135 R7=1-R8 @ I7=-I8
2140 R6=-(A*I9) @ I6=A*R9
2145 R1=R6+R7 @ I1=I6+I7
2150 GOSUB 3120
2155 FRHO=MAG-Q
2160 RETURN
3000 REM **** Z1*Z2 ****
3005 R=R1*R2-I2*I1
3010 I=R1*I2+I1*R2
3015 RETURN
3020 REM **** Z1/Z2 ****
3025 A=SQR (R1^2+I1^2)/SQR (R2^2+I2^2)
3030 T1=ATN2 (I1,R1) @ T2=ATN2 (I2,R2)
3035 R=A*COS (T1-T2)
3040 I=A*SIN (T1-T2)
3045 RETURN
3050 REM **** Z^N ****
3055 M=SQR (R1^2+I1^2)
3060 T=ATN2 (I1,R1)
3065 R=M^N*COS (N*T)
3070 I=M^N*SIN (N*T)
3075 RETURN
3080 REM **** Z1+Z2 ****
3085 R=R1+R2
3090 I=I1+I2
3095 RETURN
3100 REM **** Z1-Z2 ****
3105 R=R1-R2
3110 I=I1-I2
3115 RETURN
3120 REM **** ABS(Z1) AS MAG ****
3125 MAG=SQR (R1^2+I1^2)
3130 RETURN
3135 END
```

```
10 REM **** FLOW OVER A BOUNDED STEP  ****
20 REM ****SET UP AND LABEL
30 PLOTTER IS 705
40 INTEGER J,K,L
50 K=1 @ L=1
60 OPTION BASE 1
70 DIM V(1500),W(1500)
80 MAT V=( 0)
90 MAT W=( 0)
100 OUTPUT 705 ;"VS";2
110 GCLEAR
120 LIMIT 0,250,0,180
130 LOCATE 7.5,135,10,95
140 FRAME
150 SCALE -.5,1,0,1
160 FXD 2,2
170 CSIZE 3
180 LAXES -.25,.25,-.5,0,1,1,2
190 MOVE .1,-.05 @ LORG 4
200 LABEL "x/h"
210 MOVE -.55,.6 @ DEG @ LDIR 90
220 LABEL "y/h"
230 MOVE .25,1.01 @ LDIR 0 @ CSIZE 6
240 LABEL "FLOW OVER A BOUNDED STEP"
250 RAD @ MOVE -.5,.5 @ DRAW 0,.5 @ DRAW 0,0 @ PEN UP
255 REM **** STREAMLINES PSI ****
260 DATA .1,.2,.3,.4,.5,.6,.7,.8,.9,10
264 READ PSI
265 K=1 @ L=1
270 IF PSI=10 THEN 335
275 GAMMA=PI *PSI
295 FOR S=-5 TO 4 STEP .05
298 RHO=EXP (S)
300 GOSUB 2000
305 GOSUB 2045
310 X=R @ Y=I @ GOSUB 1000
315 NEXT S
325 GOSUB 1050
330 GOTO 264
335 GOTO 455
455 REM **** ISOTACHS  ****
457 LINE TYPE 4,1
460 DATA .2,.5,.75,.95,1.05,1.25,1.5,1.95,2.05,2.5,3,10
465 READ Q
470 K=1 @ L=1
475 IF Q=10 THEN 515
480 FOR PHI=0 TO PI /2+PI /720 STEP PI /720
485 R=Q*COS (PHI) @ I=Q*SIN (PHI)
486 GOSUB 2045
490 X=R @ Y=I
492 GOSUB 1000
495 NEXT PHI
500 GOSUB 1050
```

```
510 GOTO 465
515 GOTO 760
760 REM **** FINAL LABELING ****
765 CSIZE 4 @ LORG 1 @ DEG
770 MOVE -.45,.25
775 LABEL "STREAMLINES,PSI"
780 MOVE -.45,.15
785 LABEL "ISOTACHS,Q"
790 X=-.485 @ CSIZE 2.5 @ LORG 1
795 LDIR 0
800 DATA .505,0,.555,.1,.605,.2,.655,.3,.705,.4,.755,.5
805 DATA .805,.6,.855,.7,.905,.8,.955,.9
807 FOR J=1 TO 10
810 READ Y,PSI
815 MOVE X,Y
820 LABEL USING 825 ; "PSI=",PSI
825 IMAGE K,K
830 NEXT J
835 MOVE .75,.99 @ LORG 3
840 LABEL "PSI=1"
855 LDIR 0
870 LORG 8
885 X=-.005
890 DATA .09,.2,.215,.5,.295,.75,.34,.95,.362,1.05
891 DATA .391,1.25,.42,1.5
895 FOR J=1 TO 7
900 READ Y,Q
905 MOVE X,Y
910 LABEL USING 825 ; "Q=",Q
920 NEXT J
925 Y=.495 @ LDIR 90 @ LORG 8
930 DATA -.035,3,-.08,2.5,-.36,2.05
935 FOR J=1 TO 3
937 IF J>2 THEN CSIZE 2.5 ELSE CSIZE 2
940 READ X,Q
945 MOVE X,Y
950 LABEL USING 825 ; "Q=",Q
955 NEXT J
960 MOVE -.3,.925 @ LDIR 0 @ LORG 2 @ LABEL "Q=1.95"
965 END
1000 REM **** STORES X,Y, AND COUNTS
1010 V(K)=X @ W(K)=Y
1020 L=K
1030 K=K+1
1040 RETURN
1050 REM **** PLOTS POINTS WITHOUT REFLECTION ****
1060 K=1
1070 MOVE V(K),W(K)
1080 FOR K=1 TO L
1090 PLOT V(K),W(K),-1
1100 NEXT K
1110 PEN UP
1120 RETURN
1130 REM **** PLOTS POINTS X AXIS REFLECTED ****
```

```
1140 K=1
1150 MOVE -V(K),W(K)
1160 FOR K=1 TO L
1170 PLOT -V(K),W(K),-1
1180 NEXT K
1190 PEN UP
1200 RETURN
1210 END
1300 REM **** PLOTS POINTS WITH BOTH X & Y REFLECTED ****
1305 K=1
1310 MOVE -V(K),-W(K)
1315 FOR K=1 TO L
1320 PLOT -V(K),-W(K),-1
1325 NEXT K
1330 PEN UP
1335 RETURN
1340 REM **** PLOTS POINTS WITH Y REFLECTED ****
1345 K=1
1350 MOVE V(K),-W(K)
1355 FOR K=1 TO L
1360 PLOT V(K),-W(K),-1
1365 NEXT K
1370 PEN UP
1375 RETURN
2000 REM **** CALCULATES t ****
2005 R1=RHO*COS (GAMMA)-4
2010 I1=RHO*SIN (GAMMA)
2015 R2=RHO*COS (GAMMA)-1
2020 I2=I1
2025 GOSUB 3000
2030 N=2 @ R1=R @ I1=I
2035 GOSUB 3030
2040 RETURN
2045 REM **** CALCULATES z/h ****
2050 R1=R+1 @ I1=I
2055 R2=1-R @ I2=-I
2060 R3=R+2 @ I3=I
2065 R4=2-R @ I4=-I
2070 GOSUB 3000
2075 R1=R @ I1=I
2080 GOSUB 3070
2085 R8=R @ I8=I
2090 R1=R3 @ I1=I3
2095 R2=R4 @ I2=I4
2100 GOSUB 3000
2105 R1=R @ I1=I
2110 GOSUB 3070
2115 R7=-(R/2) @ I7=-(I/2)
2120 R=(R7+R8)/PI  @ I=(I7+I8)/PI
2125 RETURN
3000 REM **** Z1/Z2 ****
3005 A=SQR (R1^2+I1^2)/SQR (R2^2+I2^2)
3010 T1=ATN2 (I1,R1) @ T2=ATN2 (I2,R2)
3015 R=A*COS (T1-T2)
```

```
3020 I=A*SIN (T1-T2)
3025 RETURN
3030 REM **** PRINCIPAL VALUE OF Z^(1/N) ****
3035 M=SQR (R1^2+I1^2)
3040 T=ATN2 (I1,R1)
3045 R2=COS (T/N)
3050 I2=SIN (T/N)
3055 R=M^(1/N)*R2
3060 I=M^(1/N)*I2
3065 RETURN
3070 REM **** LOG(Z) ****
3075 R=LOG (SQR (R1^2+I1^2))
3080 I=ATN2 (I1,R1)
3085 RETURN
3090 END
```

```
10 REM **** FLOW OVER A FLAT PLATE WITH CIRCULATION ***
15 REM **** ALPHA=15 ****
20 REM ****SET UP AND LABEL
30 PLOTTER IS 705
40 INTEGER K,L,J,P,N,JJ
50 K=1 @ L=1
60 OPTION BASE 1
70 DIM V(2500),W(2500)
80 MAT V=( 0)
90 MAT W=( 0)
100 OUTPUT 705 ;"VS";2
110 GCLEAR
120 LIMIT 0,250,0,180
130 LOCATE 7.5,135,10,95
140 FRAME
150 SCALE -6,6,-4,4
160 FXD 0
170 CSIZE 3
180 LAXES -1,1,-6,-4,1,1,2
190 MOVE 0,-4.6 @ LORG 4
200 LABEL "x/c"
210 MOVE -6.4,0 @ DEG @ LDIR 90
220 LABEL "y/c"
230 MOVE 0,4.1 @ LDIR 0 @ CSIZE 6
240 LABEL "THE FLAT PLATE AIRFOIL; ALPHA=15"
250 RAD
251 PEN 2
252 MOVE -2,0 @ DRAW 2,0 @ PEN UP
253 PEN 1
260 REM *** DRAW PSI=0 ***
262 P=0 @ K=1 @ L=1
265 ALP=PI /12
270 RHO=1 @ DRHO=.05
275 X=2 @ Y=0 @ GOSUB 1000
280 FOR J=1 TO 150
300 RHO=RHO+DRHO
305 GOSUB 1400
315 GOSUB 2000
320 X=R @ Y=I @ GOSUB 1000
322 IF ABS (X)>6 THEN 330
325 NEXT J
330 GOSUB 1050
335 P=P+1 @ K=1 @ L=1
350 IF P=1 THEN 351 ELSE 360
351 X=-(2*COS (2*ALP)) @ Y=0 @ GOSUB 1000
352 RHO=1 @ DRHO=.05
355 GOTO 280
360 GOTO 385
385 REM **** STREAMLINES PSI>0 ****
387 P=0
390 DATA .25,.5,.75,1,1.5,2,2.5,3,10
395 READ PSI
400 K=1 @ L=1 @ ALP=PI /12
```

```
405 IF PSI=10 THEN 455
410 RHO=1.5 @ NU=ALP+PI /2 @ GOSUB 1500 @ GOSUB 2000
412 X=R @ Y=I @ GOSUB 1000
415 DRHO=RHO/100
420 FOR J=1 TO 1000
425 RHO=RHO+DRHO @ GOSUB 1550 @ GOSUB 2000
430 X=R @ Y=I
435 GOSUB 1000
437 IF ABS (X)>6 THEN 445
440 NEXT J
445 GOSUB 1050
450 GOTO 395
455 P=P+1
460 IF P=1 THEN 462 ELSE 470
462 RESTORE 390
465 GOTO 395
470 REM **** PSI<0 ****
471 P=0
475 DATA -.25,-.5,-.75,-1,-1.5,-2,10
480 READ PSI
485 K=1 @ L=1 @ ALP=PI /12
486 IF PSI=10 THEN 525
487 RHO=1.5 @ NU=4.9742 @ DNU=PI /360
490 FOR J=1 TO 178
495 GOSUB 1500 @ GOSUB 2000
498 X=R @ Y=I @ GOSUB 1000
499 IF ABS (X)>6 THEN 510
500 IF P=0 THEN NU=NU+DNU ELSE NU=NU-DNU
505 NEXT J
510 GOSUB 1050
515 GOTO 480
525 P=P+1
530 IF P=1 THEN 535 ELSE 705
535 RESTORE 475
540 GOTO 480
544 REM **** Q<>1 ****
545 LINE TYPE 4,1
563 ! DATA .9,3.41,5.78,1.1,.9599,3.39,.9659,3.406,6.2657
564 ! DATA 10,10,10
565 READ Q,NU1,NU2
567 K=1 @ L=1
568 IF Q=10 THEN 583
569 DNU=(NU2-NU1)/300
571 GOSUB 1800
573 FOR J=1 TO 299
575 GOSUB 1850
577 NEXT J
579 GOSUB 1900
581 GOTO 565
583 ! DATA 1.25,1.658,3.375,.75,3.44,4.88,10,10,10
585 READ Q,NU1,NU2
587 K=1 @ L=1
589 IF Q=10 THEN 605
591 DNU=(NU2-NU1)/200
```

```
593 GOSUB 1800
595 FOR J=1 TO 199
597 GOSUB 1850
599 NEXT J
601 GOSUB 1900
603 GOTO 585
605 ! DATA 1.5,2.234,3.351,2,2.652,3.316
607 ! DATA .5,3.49,4.153,.25,3.56,3.83,10,10,10
609 READ Q,NU1,NU2
611 K=1 @ L=1
613 IF Q=10 THEN 629
615 DNU=(NU2-NU1)/100
617 GOSUB 1800
619 FOR J=1 TO 99
621 GOSUB 1850
623 NEXT J
625 GOSUB 1900
627 GOTO 609
629 GOTO 640
640 REM **** Q=1 ****
645 ! DATA 3.4,.26179,10
650 READ NU
655 K=1 @ L=1 @ RHO=1 @ DRHO=.05 @ Q=1
660 IF NU=10 THEN 800
665 GOSUB 2000
670 X=R @ Y=I @ GOSUB 1000
675 IF ABS (X)>6 THEN 695
680 IF ABS (Y)>4 THEN 695
685 RHO=RHO+DRHO
690 GOTO 665
695 GOSUB 1050
700 GOTO 650
705 REM **** FINAL LABELING ****
710 LORG 4 @ CSIZE 4 @ DEG @ LDIR 0
715 MOVE -3.5,3 @ LABEL "STREAMLINES, PSI"
720 LDIR 10 @ CSIZE 2.5 @ LORG 1 @ X=4.5
725 DATA 0,.35,.25,.6,.5,.86,.75,1.11,1,1.38,1.5,1.89,2,2.39
730 DATA 3,3.39,-.25,.1,-.5,-.15,-.75,-.4,-1,-.67,-1.5,-1.2
731 DATA 2.5,2.89,-2,-1.74
735 FOR J=1 TO 15
740 READ PSI,Y
745 MOVE X,Y
750 LABEL USING 755 ; "PSI=",PSI
755 IMAGE K,K
760 NEXT J
765 PLIST @ END
800 REM **** FINAL LABELING ISOTACHS ****
801 CSIZE 2.5 @ DEG @ LDIR 90 @ LORG 1
802 DATA 1.1,1.05,1.25,-.1,1.5,-1.18,2,-1.69,.25,-1.49
803 DATA .5,-.99,.75,.4,.9,1.69
805 FOR J=1 TO 8
806 IF J<= 4 THEN Y=.05 ELSE Y=-.05
810 READ Q,X
811 IF J>1 THEN LORG 3
```

```
812 IF J>4 THEN LORG 9
813 IF J=5 THEN CSIZE 2 ELSE CSIZE 2.5
814 IF J=8 THEN LORG 7
815 MOVE X,Y
820 LABEL USING 825 ; "Q=",Q
825 IMAGE K,K
830 NEXT J
835 LDIR 0 @ MOVE 3.9,1 @ LORG 7
840 LABEL "Q=1"
855 MOVE 4.88,-.8 @ LDIR -28
860 LORG 4 @ LABEL "Q=COS(ALPHA)"
865 LDIR 0
870 LORG 7
875 MOVE -4.65,-1
880 LABEL "Q=1"
885 LORG 4 @ LDIR 55
890 MOVE -5.3,-3 @ LABEL "Q=COS(ALPHA)"
895 LDIR 22 @ LORG 1
900 MOVE -5.7,-2.3 @ LABEL "PSI=0"
905 LDIR 0 @ CSIZE 4 @ MOVE 3,2.5 @ LABEL "ISOTACHS,Q"
910 END
915 IMAGE K,K
920 NEXT J
925 LDIR 52
930 DATA 1.19,-.2,1.5,1.41,-.37,1.25,1.65,-.57,1.15
931 DATA 2.48,-1.22,1.05,1.9,-.75,1.1
935 FOR J=1 TO 5
940 READ X,Y,Q
945 MOVE X,Y
950 LABEL USING 915 ; "Q=",Q
955 NEXT J
960 END
1000 REM **** STORES X,Y, AND COUNTS
1010 V(K)=X @ W(K)=Y
1020 L=K
1030 K=K+1
1040 RETURN
1050 REM **** PLOTS POINTS WITHOUT REFLECTION ****
1060 K=1
1070 MOVE V(K),W(K)
1080 FOR K=1 TO L
1090 PLOT V(K),W(K),-1
1100 NEXT K
1110 PEN UP
1120 RETURN
1130 REM **** PLOTS POINTS X AXIS REFLECTED ****
1140 K=1
1150 MOVE -V(K),W(K)
1160 FOR K=1 TO L
1170 PLOT -V(K),W(K),-1
1180 NEXT K
1190 PEN UP
1200 RETURN
1210 END
```

```
1300 REM **** PLOTS POINTS WITH BOTH X & Y REFLECTED ****
1305 K=1
1310 MOVE -V(K),-W(K)
1315 FOR K=1 TO L
1320 PLOT -V(K),-W(K),-1
1325 NEXT K
1330 PEN UP
1335 RETURN
1340 REM **** PLOTS POINTS WITH Y REFLECTED ****
1345 K=1
1350 MOVE V(K),-W(K)
1355 FOR K=1 TO L
1360 PLOT V(K),-W(K),-1
1365 NEXT K
1370 PEN UP
1375 RETURN
1400 REM **** RETURNS NU FOR RHO ON PSI=0 ****
1405 A1=-(2*SIN (ALP)*RHO*LOG (RHO))
1410 B1=RHO^2-1
1415 C1=-(A1/B1)
1420 IF P=0 THEN NU=ALP-ASN (C1)
1421 IF P=1 THEN NU=PI +ALP+ASN (C1)
1425 RETURN
1500 REM **** RHO ON AN SL FOR ALP,PSI&NU ****
1505 A1=SIN (NU-ALP)
1510 B1=SIN (ALP)
1515 FRHO=(RHO^2-1)*A1-RHO*PSI+2*B1*RHO*LOG (RHO)
1520 FPRHO=2*RHO*A1-PSI+2*B1*(1+LOG (RHO))
1525 RT=FRHO/FPRHO
1530 IF ABS (RT)<.0001 THEN 1545
1535 RHO=RHO-RT
1540 GOTO 1515
1545 RETURN
1550 REM **** RETURNS NU FOR RHO ON AN SL ****
1555 A1=PSI-2*SIN (ALP)*LOG (RHO)
1560 B1=RHO-1/RHO
1565 C1=A1/B1
1567 IF P=0 THEN NU=ALP+ASN (C1)
1570 IF P=1 THEN NU=PI +ALP-ASN (C1)
1575 RETURN
1580 REM **** RETURNS RHO ON AN ISOTACH FOR Q & NU ****
1585 A=1-Q^2
1590 B=Q^2*COS (2*NU)-COS (2*(NU-ALP))
1595 C=4*SIN (ALP)*SIN (NU-ALP)
1597 AA=(2*RHO*SIN (ALP))^2
1600 FRHO=A*(RHO^4+1)+2*B*RHO^2+C*(RHO^3+RHO)+AA
1605 FPRHO=4*A*RHO^3+4*B*RHO+(3*RHO^2+1)*C+8*RHO*SIN (ALP)^2
1610 RT=FRHO/FPRHO
1615 IF ABS (RT)<.0001 THEN 1630
1620 RHO=RHO-RT
1625 GOTO 1600
1630 RETURN
1650 REM **** NU ON Q=1 FOR RHO ****
1655 A1=(RHO^3+RHO)*4*SIN (ALP)
```

```
1660 B1=(2*RHO*SIN (ALP))^2
1665 A=(2*RHO)^2
1670 B=A1
1675 C=B1-2*RHO^2+1
1680 D=SQR (B^2-4*A*C)
1685 RT1=(-B+D)/(2*A)
1690 RT2=(-B-D)/(2*A)
1695 IF P=0 THEN NU=ASN (RT1)+ALP
1700 IF P=1 THEN NU=PI -ASN (RT1)+ALP
1705 IF P=2 THEN NU=ASN (RT2)+ALP
1710 IF P=3 THEN NU=-PI -ASN (RT2)+ALP
1715 RETURN
1720 REM **** RETURNS RHO FOR NU & Q=1 ****
1725 A=4*SIN (ALP)*SIN (NU-ALP)
1730 B=4*SIN (ALP)^2-2*COS (2*(NU-ALP))
1735 FRHO=(RHO^3+RHO)*A+B*RHO^2+1
1740 FPRHO=A*(3*RHO^2+1)+2*RHO*B
1745 RT=FRHO/FPRHO
1750 IF ABS (RT)<.0001 THEN 1765
1755 RHO=RHO-RT
1760 GOTO 1735
1765 RETURN
1800 REM **** SUB OF 544 ****
1805 RHO=1 @ NU=NU1
1810 GOSUB 2000
1815 X=R @ Y=I
1820 GOSUB 1000
1825 RETURN
1850 REM **** SUB OF 544 ****
1855 NU=NU+DNU
1860 GOSUB 1580
1865 GOSUB 2000
1870 X=R @ Y=-I
1875 GOSUB 1000
1880 RETURN
1900 REM **** SUB OF 544 ****
1905 NU=NU2 @ RHO=1
1910 GOSUB 2000
1915 X=R @ Y=0
1920 GOSUB 1000
1925 GOSUB 1050
1930 RETURN
2000 REM **** ZETA AS RHO&NU TO Z AS R&I - JOUKOWSKI ****
2005 R8=RHO*COS (NU)
2010 I8=RHO*SIN (NU)
2015 N=-1
2020 R1=R8 @ I1=I8
2025 GOSUB 3000
2030 R=R+R8
2035 I=I+I8
2040 RETURN
3000 REM **** Z^N ****
3005 M=SQR (R1^2+I1^2)
3010 T=ATN2 (I1,R1)
```

```
3015 R=M^N*COS (N*T)
3020 I=M^N*SIN (N*T)
3025 RETURN
3030 END
```

```
10 REM **** SYMMETRIC JOUKOWSKI FOIL,ALPHA=15 ****
20 REM ****SET UP AND LABEL
30 PLOTTER IS 705
40 INTEGER K,L,J,P,N
50 K=1 @ L=1
60 OPTION BASE 1
70 DIM V(1500),W(1500)
80 MAT V=( 0)
90 MAT W=( 0)
100 OUTPUT 705 ;"VS";2
110 GCLEAR
120 LIMIT 0,250,0,180
130 LOCATE 7.5,135,10,95
140 FRAME
150 SCALE -6,6,-4,4
160 FXD 0
170 CSIZE 3
180 LAXES -1,1,-6,-4,1,1,2
190 MOVE 0,-4.6 @ LORG 4
200 LABEL "x/c"
210 MOVE -6.4,0 @ DEG @ LDIR 90
220 LABEL "y/c"
230 MOVE 0,4.1 @ LDIR 0 @ CSIZE 6
240 LABEL "THE SYMMETRIC JOUKOWSKI AIRFOIL"
250 RAD @ ALP=PI /12
253 G=.77*.12 ! THE 12% THICKNESS
260 REM *** DRAW THE AEROFOIL ****
262 P=0 @ K=1 @ L=1 @ PEN 2
265 FOR NU=0 TO PI +PI /180 STEP PI /180
270 X=2*COS (NU)
275 Y=2*G*(1-COS (NU))*SIN (NU)
280 GOSUB 1000
300 NEXT NU
305 GOSUB 1050
315 GOSUB 1340
317 PEN 1 @ GOTO 525
320 REM **** DRAW PSI > 0 ****
321 P=0
322 ! DATA 1,-.25,-.5,-.75,-1,-1.5,-2,1.5,2,2.5,3,10
325 READ PSI
330 K=1 @ L=1
335 IF PSI=10 THEN 400
340 RHO=4
350 IF PSI>0 THEN NU=PI /2 ELSE NU=3*PI /2
351 IF P=0 THEN DNU=PI /90 ELSE DNU=-(PI /90)
352 GOSUB 1400 @ GOSUB 2000
355 X=R @ Y=I @ GOSUB 1000
360 IF ABS (X)>6 THEN 390
361 IF ABS (Y)>4 THEN 390
385 NU=NU+DNU
387 GOTO 352
390 GOSUB 1050
395 GOTO 325
```

```
400 P=P+1
405 IF P=1 THEN 410 ELSE 415
410 RESTORE 322
412 GOTO 325
415 REM **** PSI=0 ****
420 ! DATA .0175,3.6652,10
425 READ NU
427 IF NU=10 THEN 480
430 K=1 @ L=1 @ DRHO=.05 @ PSI=0
435 RHO=1+G*(1-COS (NU))
440 GOSUB 2000 @ X=R @ Y=I @ GOSUB 1000
445 RHO=RHO+DRHO
450 GOSUB 1455
455 GOSUB 2000 @ X=R @ Y=I @ GOSUB 1000
460 IF ABS (X)>6 THEN 470
465 GOTO 445
470 GOSUB 1050
475 GOTO 425
480 REM **** PSI>0 ****
482 P=0
484 ! DATA .25,3.54,.35,.5,3.54,.35,.75,3.54,.35,10,10,10
486 READ PSI,NU2,NU3
488 K=1 @ L=1 @ ALP=PI /12
490 IF PSI=10 THEN 522
492 RHO=7 @ NU=NU2 @ GOSUB 1455
494 NU2=NU @ GOSUB 1250 @ RHO2=RHO
496 RHO=1.5 @ NU=ALP+PI /2 @ GOSUB 1400
498 RHO1=RHO @ NU1=NU @ DRHO=(RHO2-RHO1)/500
500 RHO=RHO2 @ NU=NU2
502 GOSUB 1275
504 RHO=RHO1 @ NU=NU1 @ GOSUB 1250
506 GOSUB 1050
508 K=1 @ L=1
510 RHO=7 @ NU=NU3 @ GOSUB 1455
512 NU3=NU @ GOSUB 1250
514 GOSUB 1275
516 RHO=RHO1 @ NU=NU1 @ GOSUB 1250
518 GOSUB 1050
520 GOTO 486
522 GOTO 867
525 REM **** Q<>1 ****
529 DATA 1.25,1.459,3.3586,1.5,1.9373,3.328,2,2.479,3.281
530 DATA 10,10,10
531 LINE TYPE 4,1
533 READ Q,NU1,NU2
535 K=1 @ L=1
537 IF Q=10 THEN 559
539 DNU=(NU2-NU1)/100 @ NU=NU1
541 RHO=1+G*(1-COS (NU))
543 GOSUB 1250
545 FOR J=1 TO 100
547 NU=NU+DNU
549 GOSUB 1580 @ DISP J
551 GOSUB 1250
```

```
553 NEXT J
555 GOSUB 1050
557 GOTO 533
559 DATA .5,3.477,3.961,1.05,.89,3.382,1.1,1.0646,3.3745
560 DATA 10,10,10
561 READ Q,NU1,NU2
563 K=1 @ L=1
565 IF Q=10 THEN 605
567 NU=(NU1+NU2)/2 @ NU3=NU
568 IF Q=.5 THEN RHO=2 ELSE RHO=5
569 GOSUB 1580 @ RHOM=RHO @ PRINT RHOM
571 RHOB=1+G*(1-COS (NU1)) @ DRHO=(RHOM-RHOB)/100
573 RHO=RHOB @ NU=NU1 @ GOSUB 1250
575 FOR J=1 TO 99
577 RHO=RHO+DRHO @ GOSUB 1650 @ GOSUB 1250 @ DISP J
579 IF ABS (Y)>4 THEN 583
581 NEXT J
582 RHO=RHOM @ NU=NU3 @ GOSUB 1250
583 GOSUB 1050 @ K=1 @ L=1
585 RHOB=1+G*(1-COS (NU2)) @ DRHO=(RHOM-RHOB)/100
587 RHO=RHOB @ NU=NU2 @ GOSUB 1250
589 FOR J=1 TO 99
591 RHO=RHO+DRHO @ GOSUB 1650 @ GOSUB 1250 @ DISP J
593 IF ABS (Y)>4 THEN 597
595 NEXT J
596 RHO=RHOM @ NU=NU3 @ GOSUB 1250
597 GOSUB 1050
599 GOTO 561
605 REM **** Q=1 ****
607 Q=1 @ NU=.6951
609 K=1 @ L=1
611 RHO=1+G*(1-COS (NU)) @ DNU=-(PI /360)
613 GOSUB 1250
615 NU=NU+DNU
617 GOSUB 1580
619 GOSUB 1250
621 IF ABS (X)>6 THEN 625
623 GOTO 615
625 GOSUB 1050
626 K=1 @ L=1
627 NU=3.387
629 RHO=1+G*(1-COS (NU)) @ DRHO=.01
631 GOSUB 1250
633 RHO=RHO+DRHO
635 GOSUB 1650
637 GOSUB 1250
639 IF ABS (X)>6 THEN 643
641 GOTO 633
643 GOSUB 1050
645 GOTO 730
730 REM **** Q<1 ****
731 DATA .884246,3.405,6.283185,.95,3.3968,6.7369
732 DATA .85,3.412,5.864
735 DATA .75,3.427,4.5,10,10,10
```

```
896 MOVE 3.95,1.4 @ LABEL "Q=1"
897 MOVE -5.31,-.95 @ LABEL "Q=1"
898 LDIR 0 @ LORG 1
899 DATA -4.4,-2.5,.95,-5.1,.35,1.05,-4.46,1.3,1.1,1.45,2.05
900 DATA 1.1,2.5,1.9,1.05,4.08,-.69,.95,2.1,-1.5,.8843
901 FOR J=1 TO 7
902 READ X,Y,Q
903 MOVE X,Y
904 LABEL USING 893 ; "Q=",Q
905 NEXT J
906 END
1000 REM **** STORES X,Y, AND COUNTS
1010 V(K)=X @ W(K)=Y
1020 L=K
1030 K=K+1
1040 RETURN
1050 REM **** PLOTS POINTS WITHOUT REFLECTION ****
1060 K=1
1070 MOVE V(K),W(K)
1080 FOR K=1 TO L
1090 PLOT V(K),W(K),-1
1100 NEXT K
1110 PEN UP
1120 RETURN
1130 REM **** PLOTS POINTS X AXIS REFLECTED ****
1140 K=1
1150 MOVE -V(K),W(K)
1160 FOR K=1 TO L
1170 PLOT -V(K),W(K),-1
1180 NEXT K
1190 PEN UP
1200 RETURN
1210 END
1250 REM **** JOUK ZETA TO Z AND STORE ****
1255 GOSUB 2000
1260 X=R @ Y=I
1265 GOSUB 1000
1270 RETURN
1275 REM **** DECREMENT & STORE RHO & NU ON A SL ****
1280 FOR J=1 TO 499
1285 RHO=RHO-DRHO
1290 GOSUB 1455
1295 GOSUB 1250
1296 NEXT J
1297 RETURN
1300 REM **** PLOTS POINTS WITH BOTH X & Y REFLECTED ****
1305 K=1
1310 MOVE -V(K),-W(K)
1315 FOR K=1 TO L
1320 PLOT -V(K),-W(K),-1
1325 NEXT K
1330 PEN UP
1335 RETURN
1340 REM **** PLOTS POINTS WITH Y REFLECTED ****
```

```
1345 K=1
1350 MOVE V(K),-W(K)
1355 FOR K=1 TO L
1360 PLOT V(K),-W(K),-1
1365 NEXT K
1370 PEN UP
1375 RETURN
1400 REM **** RETURNS RHO FOR NU & PSI ON A SL ****
1402 A=(1+G)*SIN (ALP)
1405 F1=(G^2+2*G*RHO*COS (NU)+RHO^2)/(1+G)^2
1410 F2=RHO*SIN (NU-ALP)-G*SIN (ALP)
1415 FRHO=F2*(F1-1)+A*F1*LOG (F1)-PSI*F1
1420 FP1=(2*RHO+2*G*COS (NU))/(1+G)^2
1422 AA=PSI*FP1
1425 FPRHO=F2*FP1+(F1-1)*SIN (NU-ALP)+A*(FP1+FP1*LOG (F1))-AA
1430 RT=FRHO/FPRHO
1435 IF ABS (RT)<.0001 THEN 1450
1440 RHO=RHO-RT
1445 GOTO 1405
1450 RETURN
1455 REM **** RETURNS NU FOR RHO&PSI ON A SL ****
1460 A=(1+G)*SIN (ALP)
1465 F1=(G^2+2*G*RHO*COS (NU)+RHO^2)/(1+G)^2
1470 F2=RHO*SIN (NU-ALP)-G*SIN (ALP)
1475 FP1=-(2*G*RHO*SIN (NU)/(1+G)^2)
1480 FP2=RHO*COS (NU-ALP)
1485 FU=F2*(F1-1)+A*F1*LOG (F1)-PSI*F1
1490 FPNU=F2*FP1+(F1-1)*FP2+A*(FP1+FP1*LOG (F1))-PSI*FP1
1495 RT=FU/FPNU
1500 IF ABS (RT)<.0001 THEN 1515
1505 NU=NU-RT
1510 GOTO 1465
1515 RETURN
1580 REM **** RETURNS RHO FOR Q & NU --N & R ****
1585 DRHO=.05 @ E=.0001
1590 GOSUB 2100
1595 F1=FRHO
1600 RHO=RHO+DRHO
1605 GOSUB 2100
1610 F2=FRHO
1615 F3=(F2-F1)/DRHO
1617 IF F3=0 THEN 1640
1620 RT=F1/F3
1625 IF ABS (RT)<E THEN 1640
1630 RHO=RHO-DRHO-RT
1635 GOTO 1590
1640 RHO=RHO-DRHO
1645 RETURN
1650 REM **** RETURNS NU FOR Q&RHO - NR ****
1655 DNU=.015 @ E=.0001
1660 GOSUB 2100
1665 F1=FRHO
1670 NU=NU+DNU
1675 GOSUB 2100
```

```
1680 F2=FRHO
1685 F3=(F2-F1)/DNU
1690 IF F3=0 THEN 1715
1695 RT=F1/F3
1700 IF ABS (RT)<.0001 THEN 1715
1705 NU=NU-DNU-RT
1710 GOTO 1660
1715 NU=NU-DNU
1720 RETURN
2000 REM **** ZETA AS RHO&NU TO Z AS R&I - JOUKOWSKI ****
2005 R8=RHO*COS (NU)
2010 I8=RHO*SIN (NU)
2015 N=-1
2020 R1=R8 @ I1=I8
2025 GOSUB 3050
2030 R=R+R8
2035 I=I+I8
2040 RETURN
2100 REM **** RETURNS Q FOR NU ON RHO=1 ****
2101 A=(1+G)^2
2102 B=2*(1+G)*SIN (ALP)
2103 C=COS (ALP) @ D=SIN (ALP)
2105 R10=RHO*COS (NU) @ I10=RHO*SIN (NU)
2110 N=2 @ R1=R10 @ I1=I10
2115 GOSUB 3050 @ R9=R @ I9=I
2120 R1=R9-1 @ I1=I9 @ N=-1
2125 GOSUB 3050 @ R1=R @ I1=I
2130 R2=R9 @ I2=I9 @ GOSUB 3000
2135 R8=R @ I8=I
2140 R1=R10+G @ I1=I10 @ N=-2
2145 GOSUB 3050 @ R1=A*R @ I1=A*I
2150 R2=-C @ I2=-D @ GOSUB 3000
2155 R7=R @ I7=I
2160 R1=R10+G @ I1=I10 @ N=-1
2165 GOSUB 3050 @ R5=B*R @ I5=B*I
2170 R6=-I5 @ I6=R5
2175 R5=C @ I5=-D
2180 R1=R5+R6+R7 @ I1=I5+I6+I7
2185 R2=R8 @ I2=I8
2190 GOSUB 3000 @ R1=R @ I1=I
2195 GOSUB 3120
2200 FRHO=MAG-Q
2205 RETURN
3000 REM **** Z1*Z2 ****
3005 R=R1*R2-I2*I1
3010 I=R1*I2+I1*R2
3015 RETURN
3020 REM **** Z1/Z2 ****
3025 A=SQR (R1^2+I1^2)/SQR (R2^2+I2^2)
3030 T1=ATN2 (I1,R1) @ T2=ATN2 (I2,R2)
3035 R=A*COS (T1-T2)
3040 I=A*SIN (T1-T2)
3045 RETURN
3050 REM **** Z^N ****
```

```
3055 M=SQR (R1^2+I1^2)
3060 T=ATN2 (I1,R1)
3065 R=M^N*COS (N*T)
3070 I=M^N*SIN (N*T)
3075 RETURN
3080 REM **** Z1+Z2 ****
3085 R=R1+R2
3090 I=I1+I2
3095 RETURN
3100 REM **** Z1-Z2 ****
3105 R=R1-R2
3110 I=I1-I2
3115 RETURN
3120 REM **** ABS(Z1) AS MAG ****
3125 MAG=SQR (R1^2+I1^2)
3130 RETURN
3135 END
```

```
10 REM **** BORDA'S MOUTHPIECE  ****
20 REM **** SET UP AND LABEL ****
30 PLOTTER IS 705
40 INTEGER K,L,J,P,N
50 K=1 @ L=1
60 OPTION BASE 1
70 DIM V(1500),W(1500)
80 MAT V=( 0)
90 MAT W=( 0)
100 OUTPUT 705 ;"VS";2
110 GCLEAR
120 LIMIT 0,250,0,180
130 LOCATE 7.5,135,10,95
140 FRAME
150 SCALE -3,3,-2,2
160 FXD 0
170 CSIZE 3
180 LAXES -1,1,-3,-2,1,1,2
190 MOVE 0,-2.3 @ LORG 4
200 LABEL "x/a"
210 MOVE -3.2,0 @ DEG @ LDIR 90
220 LABEL "y/a"
230 MOVE 0,2.05 @ LDIR 0 @ CSIZE 6
240 LABEL "BORDA'S MOUTHPIECE"
250 RAD
253 PEN 2 @ MOVE -3,1 @ DRAW 0,1 @ PEN UP
254 MOVE -3,-1 @ DRAW 0,-1 @ PEN UP @ PEN 1
260 REM *** DRAW THE ORIFICE AND THE FREE STREAMLINES ****
262 K=1 @ L=1
265 FOR TH=-(PI /180) TO -PI  STEP -(PI /180)
270 X=(SIN (TH/2)^2-LOG (SEC (TH/2)))/PI
275 Y=(TH-SIN (TH))/(2*PI )+1
280 GOSUB 1000
300 NEXT TH
302 X=-3 @ Y=.5 @ GOSUB 1000
305 GOSUB 1340
306 LINE TYPE 4,1 @ GOSUB 1050 @ LINE TYPE 1
307 MOVE -3,0 @ DRAW 3,0 @ PEN UP
320 REM **** DRAW PSI > 0 ****
322 DATA .1,.2,.3,.4,10
325 READ PSI
330 K=1 @ L=1 @ J=0
335 IF PSI=10 THEN 387
340 PHI=PI *PSI
350 FOR X1=-12 TO 5.5 STEP .1
351 RHO=EXP (X1)
352 GOSUB 2000
355 GOSUB 1000 @ J=J+1 @ DISP J
360 NEXT X1
361 GOSUB 1050
385 GOTO 325
387 GOTO 525
525 REM **** Q<=1 ****
```

```
530 DATA .95,-.8312,.99,-1.3634,.8,-.3038,.7,-.1038,.6,.0747
531 DATA .5,.2568,.9,-.584,.4,.4654
532 DATA .3,.7387,.2,1.1755,.1,2.2294,.05,4.0588,10,10
533 READ Q,XO
535 K=1 @ L=1 @ LINE TYPE 4,1
537 IF Q=10 THEN 555
539 LAM=1/Q @ GAM=0 @ DGAM=PI /90
541 FOR J=1 TO 97
543 GOSUB 2100 @ GOSUB 2000 @ GOSUB 1000 @ DISP J
544 IF J>= 90 THEN DGAM=PI /720
545 GAM=GAM+DGAM
547 NEXT J
549 X=XO @ Y=0 @ GOSUB 1000
553 GOSUB 1340
554 GOTO 533
555 GOTO 867
867 REM **** FINAL LABELING STREAMLINES ****
868 CSIZE 4 @ LORG 1 @ DEG @ LDIR 0
869 MOVE -2.5,-1.5 @ LABEL "STREAMLINES,PSI"
870 CSIZE 2.5
871 DATA -.55,-1.9,.1,.45,-1.9,.2,1.4,-1.9,.3,2.35,-1,.4
872 DATA -2.75,-1.05,0,-2.75,.95,1,2.5,-.05,.5
874 FOR J=1 TO 7
875 READ X,Y,PSI
876 IF J>3 THEN LORG 3 ELSE LORG 7
877 MOVE X,Y @ LABEL USING 880 ; "PSI=",PSI
878 IF J>6 THEN LORG 3
879 NEXT J
880 IMAGE K,K
882 REM **** FINAL LABELING ISOTACHS ****
883 CSIZE 4 @ LORG 1 @ DEG @ LDIR 0
885 MOVE 1.75,1.75 @ LABEL "ISOTACHS,Q"
886 CSIZE 2.5 @ Y=.04 @ LDIR 90
887 DATA -1.4,.99,-.87,.95,-.6,.9,-.34,.8,-.125,.7,.04,.6
888 DATA .23,.5,.43,.4,.79,.3,1.23,.2,2.3,.1
889 FOR J=1 TO 11
890 READ X,Q
891 IF J>8 THEN LORG 3 ELSE LORG 1
892 MOVE X,Y @ LABEL USING 893 ; "Q=",Q
893 IMAGE K,K
894 NEXT J
895 LDIR 90 @ LORG 1
896 MOVE -2.35,1.05 @ LABEL "Q=.05" @ LDIR 0
897 MOVE -2,.55 @ LABEL "Q=1"
899 PLIST @ END
1000 REM **** STORES X,Y, AND COUNTS
1010 V(K)=X @ W(K)=Y
1020 L=K
1030 K=K+1
1040 RETURN
1050 REM **** PLOTS POINTS WITHOUT REFLECTION ****
1060 K=1
1070 MOVE V(K),W(K)
1080 FOR K=1 TO L
```

```
1090 PLOT V(K),W(K),-1
1100 NEXT K
1110 PEN UP
1120 RETURN
1130 REM **** PLOTS POINTS X AXIS REFLECTED ****
1140 K=1
1150 MOVE -V(K),W(K)
1160 FOR K=1 TO L
1170 PLOT -V(K),W(K),-1
1180 NEXT K
1190 PEN UP
1200 RETURN
1210 END
1250 REM **** JOUK ZETA TO Z AND STORE ****
1255 GOSUB 2000
1260 X=R @ Y=I
1265 GOSUB 1000
1270 RETURN
1275 REM **** DECREMENT & STORE RHO & NU ON A SL ****
1280 FOR J=1 TO 499
1285 RHO=RHO-DRHO
1290 GOSUB 1455
1295 GOSUB 1250 @ DISP J
1296 NEXT J
1297 RETURN
1300 REM **** PLOTS POINTS WITH BOTH X & Y REFLECTED ****
1305 K=1
1310 MOVE -V(K),-W(K)
1315 FOR K=1 TO L
1320 PLOT -V(K),-W(K),-1
1325 NEXT K
1330 PEN UP
1335 RETURN
1340 REM **** PLOTS POINTS WITH Y REFLECTED ****
1345 K=1
1350 MOVE V(K),-W(K)
1355 FOR K=1 TO L
1360 PLOT V(K),-W(K),-1
1365 NEXT K
1370 PEN UP
1375 RETURN
1580 REM **** RETURNS RHO FOR Q & NU --N & R ****
1585 DRHO=.05 @ E1=.0001
1590 GOSUB 2100
1595 F1=FRHO
1600 RHO=RHO+DRHO
1605 GOSUB 2100
1610 F2=FRHO
1615 F3=(F2-F1)/DRHO
1617 IF F3=0 THEN 1640
1620 RT=F1/F3
1625 IF ABS (RT)<E1 THEN 1640
1630 RHO=RHO-DRHO-RT
1635 GOTO 1590
```

```
1640 RHO=RHO-DRHO
1645 RETURN
1650 REM **** RETURNS NU FOR Q&RHO - NR ****
1655 DNU=.015 @ E1=.0001
1660 GOSUB 2100
1665 F1=FRHO
1670 NU=NU+DNU
1675 GOSUB 2100
1680 F2=FRHO
1685 F3=(F2-F1)/DNU
1690 IF F3=0 THEN 1715
1695 RT=F1/F3
1700 IF ABS (RT)<.0001 THEN 1715
1705 NU=NU-DNU-RT
1710 GOTO 1660
1715 NU=NU-DNU
1720 RETURN
2000 REM **** t TO z TRANSFORMATION; RETURNS X,Y ****
2005 R10=RHO*COS (PHI) @ I10=RHO*SIN (PHI)
2010 R1=R10 @ I1=I10 @ N=2 @ GOSUB 3050
2015 R9=R @ I9=I
2020 R1=R9-1 @ I1=I9 @ N=2 @ GOSUB 3135
2025 R8=R @ I8=I
2030 R1=R10 @ I1=I10 @ R2=R8 @ I2=I8 @ GOSUB 3000
2035 R7=R @ I7=I
2040 R1=R10 @ I1=I10 @ GOSUB 3160
2045 R6=R @ I6=I
2050 R1=R10+R8 @ I1=I10+I8 @ GOSUB 3160
2055 R5=R @ I5=I
2060 R=(1-R9-R7+R6+R5)/PI
2065 I=-((I9+I7-I6-I5)/PI )-1
2070 X=R @ Y=I
2075 RETURN
2100 REM **** ZETA TO t TRANSFORMATION ****
2105 R5=.5*LOG (LAM) @ I5=.5*GAM
2110 R=FNCOSH(R5)*COS (I5)
2115 I=FNSINH(R5)*SIN (I5)
2120 RHO=SQR (R^2+I^2)
2125 PHI=ATN2 (I,R)
2130 RETURN
2135 DEF FNCOSH(S) = (EXP (S)+EXP (-S))/2
2140 DEF FNSINH(U) = (EXP (U)-EXP (-U))/2
3000 REM **** Z1*Z2 ****
3005 R=R1*R2-I2*I1
3010 I=R1*I2+I1*R2
3015 RETURN
3020 REM **** Z1/Z2 ****
3025 A=SQR (R1^2+I1^2)/SQR (R2^2+I2^2)
3030 T1=ATN2 (I1,R1) @ T2=ATN2 (I2,R2)
3035 R=A*COS (T1-T2)
3040 I=A*SIN (T1-T2)
3045 RETURN
3050 REM **** Z^N ****
3055 M=SQR (R1^2+I1^2)
```

```
3060 T=ATN2 (I1,R1)
3065 R=M^N*COS (N*T)
3070 I=M^N*SIN (N*T)
3075 RETURN
3080 REM **** Z1+Z2 ****
3085 R=R1+R2
3090 I=I1+I2
3095 RETURN
3100 REM **** Z1-Z2 ****
3105 R=R1-R2
3110 I=I1-I2
3115 RETURN
3120 REM **** ABS(Z1) AS MAG ****
3125 MAG=SQR (R1^2+I1^2)
3130 RETURN
3135 REM **** Z^1/N:FIRST ROOT ****
3140 M=SQR (R1^2+I1^2) @ T=ATN2 (I1,R1)
3145 RR=COS (T/N) @ II=SIN (T/N)
3150 R=M^(1/N)*RR @ I=M^(1/N)*II
3155 RETURN
3160 REM **** LOG(Z) ****
3165 R=LOG (SQR (R1^2+I1^2))
3170 I=ATN2 (I1,R1)
3175 RETURN
3180 END
```

Index